International Tendency of
SOFT FURNISHING
HOME DECORATIONISM

高迪国际 HI-DESIGN PUBLISHING 编   钱源 译

国际软装趋势——室内装饰主义

江苏科学技术出版社

图书在版编目（CIP）数据

国际软装趋势：室内装饰主义 / 高迪国际编；钱源译. -- 南京：江苏科学技术出版社，2014.4
ISBN 978-7-5537-2948-0

Ⅰ. ①国… Ⅱ. ①高… ②钱… Ⅲ. ①室内装饰设计－作品集－世界－现代 Ⅳ. ①TU238

中国版本图书馆CIP数据核字（2014）第045358号

国际软装趋势——室内装饰主义

编　　者	高迪国际 HI-DESIGN PUBLISHING
译　　者	钱　源
责 任 编 辑	刘屹立
特 约 编 辑	林　溪
出 版 发 行	凤凰出版传媒股份有限公司 江苏科学技术出版社
出版社地址	南京市湖南路1号A楼，邮编：210009
出版社网址	http://www.pspress.cn
总 经 销	天津凤凰空间文化传媒有限公司
总经销网址	http://www.ifengspace.cn
经　　销	全国新华书店
印　　刷	利丰雅高印刷（深圳）有限公司
开　　本	1 020 mm×1 440 mm　1 / 16
印　　张	22.75
字　　数	255 000
版　　次	2014年4月第1版
印　　次	2014年4月第1次印刷
标 准 书 号	ISBN 978-7-5537-2948-0
定　　价	358.00元（精）

图书如有印装质量问题，可随时向销售部调换（电话：022-87893668）。

PREFACE I 序言一

Nikki Hunt, Principal

Design Intervention i.d.

Interior design has the power to influence mood, improve function and impart a sense of wellbeing. Successful design transforms how we live and feel and improves our quality of life. A designer employs a myriad of tools in the creation process: as professionals, we know that soft furnishings can transform interiors more than any other element in our arsenal. Despite the huge investment in constructing a house, it is the final flourish that imparts a sense of style and personality. A room will look and feel cold and unfinished without the addition of well-considered soft furnishings. They do for a room what clothes do for our bodies. Through the injection of color, pattern, texture and softness they draw the eye highlighting and enhancing unique features of the building while at the same time helping to cover any architectural compromises.

At design intervention id, we use decorative accessories to add balance to a space. Where we need vertical elements, we may add a striking standing lamp to draw the eye or hang curtains high above the window frame. Subtle injection of color, through a lampshade or a simple cushion, may be all that is required to enliven a home or provide a visual link from one room to the next. The addition of a rug, particularly in open concept interiors, allows us to define areas, while at the same time adding warmth and life to a room.

Our homes should reflect our personalities, taste and lifestyle and nothing does this more effectively than soft furnishing. It is the addition of well-chosen decorative elements that allows us to personalize a space and transforms a well-executed project into a place where our clients truly feel at home.

Nikki Hunt

室内设计能够影响居住者的心情、完善室内功能并带给人幸福感。成功的室内设计能够改变我们的生活和感觉，提高我们的生活质量。一个设计师在进行室内设计的过程中会运用各种各样的工具：作为专业设计人员，我们深知软装设计对室内整体风格的转换具有至关重要的作用。除了在房屋建造中投入大量心血之外，公寓的布置对凸显其整体风格和特点起到决定性的作用。如果缺少精心陈设的软装饰，整个空间会给人以冰冷和不完整的感觉。室内装饰的过程，就像给人穿衣服一样，我们给整个室内也披上了一层软装饰的外衣。室内色彩的注入，图案、纹理和软装饰的使用，不但使整个空间引人注目，也凸显和提升了空间的独特气质，同时掩饰了建筑设计中的折中之处。

作为设计干预，我们使用装饰性配饰以增强空间平衡感。当需要垂直的设计元素时，我们会在室内摆放一个醒目的落地灯或者把窗帘高高地悬挂在窗框之上。灯罩的简单使用，巧妙地为室内注入色彩，使其充满生气，并在视觉上连接相邻的房间。在这个开放式空间中铺设一块地毯，不仅能够界定空间，还能营造出一室的温馨与舒适。

住宅应该彰显出居住者的个性、品位和生活方式，而软装设计是这些特质最为有效的载体。我们运用精心挑选的装饰元素赋予空间独特的个性，并使一个执行良好的装饰设计项目转换成住户理想的家园。

Nikki Hunt

PREFACE II 序言二

Mariangel Coghlan, President

MARIANGEL COGHLAN

Human acts imply such a variety of behavior, communication and action symbols that it is impossible to establish a complete and final hierarchy about what our essence involves.

However, it is relatively simple to find essential elements that, commonly, determine our daily actions. Amongst many situations, an inescapable influence upon our lives is the physical environment which surrounds us, and within this, the space we call "home" is fundamental. This is the reason why we can exalt the importance of interior design in our lives.

From the first traces in caves, the Eskimo igloo, the stilts on water, the teepee of Native Americans in America, and the yurta of the nomads in Central-Asian steppes; in the dwellings of paranoiac Egypt, ancient Greece, the Roman empire or pre-Hispanic cultures; up to the current minimalist residences or kit homes; in apartments or houses; made from brick, wood or concrete; dwellers have always had to invest time, thought and illusions in fixing the spaces in which they dwell. All of this with a central goal which is, finally, the task of interior design, shaping our immediate space, so that people can live better.

Just as the saying of "butterfly effect": the wind produced by the flutter of wings of a dragonfly in the Caribbean can end as a tsunami in Asia. The Mexican company MARIANGEL COGHLAN firmly believes it. The style of MARIANGEL COGHLAN is the result of a reflection on the international interdependency of interior design styles in the light of the wonderful forms, colors, and natural resources offered particularly by Mexico.

With this inspiration as its starting point, the MARIANGEL COGHLAN company is fully focused in achieving the central goal of interior design: to collaborate in shaping a better world.

Mariangel Coghlan

人类的一举一动暗示着各种各样的行为、交流和活动的符号，所以要想建立一个关于人类本质的全面、终极的等级系统几乎是不可能的。

然而，决定我们日常生活的主要元素却相对简单。在许多环境中，对生活产生不可避免的影响的一个因素便是我们身处的自然环境。而在这些自然环境中，居住的房屋可以说是最基本的。这也是室内装饰设计的重要性被反复强调的原因。

从最初洞穴中人类的足迹，爱斯基摩人的冰屋，水上住宅的桩柱，美洲印第安人居住的圆锥形帐篷，中亚大草原上游牧民族居住的圆顶帐篷，古埃及人、古希腊人、罗马帝国和前哥伦布文化时期的寓所，到当下极简主义风格的住宅和工具箱式房屋……无论是公寓还是独立式住宅，无论是使用了砖块、木材还是混凝土建材，居住者总要花费时间，去思考，去幻想他们理想中的家居环境。所有这些意识和行为为室内装饰设计设定了任务，以打造这个和我们距离最近的空间，使我们能够更好地生活。

正如蝴蝶效应那样，一只在加勒比海岛上的蜻蜓扇了扇翅膀，便能够在亚洲引起一场海啸。MARIANGEL COGHLAN，一个墨西哥设计公司，深知这个道理。其设计风格正是国际上不同的室内装饰风格相互影响和作用的反映，并以精美的外形、色彩和墨西哥当地的天然材料为依托。

在这样的设计灵感指引下，MARIANGEL COGHLAN 设计公司全力以赴，致力于实现室内装饰的核心目标：与他人合作，创建一个更加美好的世界。

Mariangel Coghlan

*Carolina Sandri,
Senior Architect*

Casa Forma

As a result of the recent changes in the global economy, we are dealing with a more price conscious customer than ever before. Ensuring timeless design with a focus on design, quality, luxury, and at the same time, great value is very important.

Timeless design is thoughtful, versatile, and timeless. By coordinating beautiful furniture pieces with accessories to enhance the design, we can achieve a balanced interior design scheme which emanates harmony and beauty. We believe in sourcing furniture for our clients that is gorgeous, high-quality, and at the same time offers good value because good design is essentially about offering our clients value in addition to the highest quality attainable and aesthetics.

Some clients will come with an open design brief that requires a lot of our input, while other clients may come with an art collection that they would like to use as the principal aspect of the interior design. Our team is versatile and enjoys working with a diverse group of clients. When clients have pieces that mean something to them, or when we use antiques in the design, we are able to give them a new context, a new life. They add a character to a project—and a quality of energy—that's unlike any other.

We are fortunate to have clients that are excited by pushing boundaries with materials and design. Our search for unique materials, artisans and craftsmanship allows us to step off the traditional design path to discover the unique, exciting and unexpected.

At our studio, we create schemes that are a result of carefully understanding our clients' needs at the same time ensuring practicality, feasibility and functionality. With these thoughts in mind over the last seven years, we have successfully completed more than 65 projects in the United Kingdom and abroad. Elegant fabrics for the drapes have to be carefully selected as they have a big impact on the look of the property. Sheep skin throw, cashmere covers and cushions add an organic, earthy touch to the design. Luxurious pure wool velvet and silk rugs add a final sophisticated touch. Sumptuous embroidered and stone wash linen fabrics add elements of glamour to a space.

Our group is always striving to deliver the dream and a bit more and for this one needs impeccable delivery. As most of our projects are turnkey services, we want to delight clients in every possible angle from the door handle to decorative items such as Baccarat candelabras. We see all details in design as an opportunity to enchant and impress.

Carolina Sandri

由于近期全球经济形势的变化，我们本次接待了一位前所未有的关注价格的客户。在保证住宅经典设计风格的同时，客户格外注重房屋的设计、品质和奢华感，另外也十分关注设计本身的价值。

所谓永恒、经典的设计就是考虑周到，例如空间的不同功能和各种设计细节。精美的家具和装饰品的搭配陈列，提升了整体设计，并构成一个散发着和谐美感的设计方案。在进行室内装饰设计的过程中，我们一直致力于采购适合客户的，华丽、高品质并且物有所值的家具，因为从本质上来讲，好的装饰设计就是要在为客户提供价值的同时，使他们能够享受高品质和富于美感的生活。

一些客户会提出一个需要我们进行大量投入的开放式设计纲要，也有一些客户希望室内艺术收藏品的陈列能够成为室内设计的一个重要方面。我们的设计团队也是这一领域的多面手，并真诚希望与不同类型的客户进行交流和合作。当客户要求在设计中加入对他们有着特殊意义的物件，或者设计团队在装饰设计过程中选择古董作为装饰物时，我们就会赋予这个物件以全新的环境和生命特征。它们的融入也增加了整个设计项目的独特性——一种活力和能量——这是每次设计所特有的。

我们有幸与为在整体设计和装饰材料的使用方面不断超越常规和极限而感到兴奋的客户一起合作。我们努力寻找与众不同的装饰材料和技艺精湛的手工艺工匠，以最先进的设计工艺打破传统设计的常规之路，另辟蹊径，去寻找一种令人兴奋和与众不同的设计风格。

我们细心地制订既能满足客户需求又要保证实用性、可行性和功能性的设计方案。秉承这样的设计理念，在过去的七年中，我们成功完成了位于英国本土和国外的至少65个设计项目。挂帘精心选用了优雅的面料，因为它对房屋整体的感观具有至关重要的影响；羊皮床罩、羊绒毯子和靠垫增加了整体设计的有机感和质朴感；奢华纯羊绒材质天鹅绒和真丝面料地毯尽显精致；华美的刺绣和石洗花纹亚麻面料的选择也使空间彰显出迷人的魅力。

我们的设计团队一直致力于传递梦想和超越梦想。因为我们大部分的设计项目都是承包服务，所以从房门扶手到室内装饰物，例如Baccarat枝状大烛台，都力求从尽可能多的角度去取悦客户。我们把每一个装饰细节设计都当作展示室内空间迷人风采和给人留下深刻记忆的宝贵机会。

Carolina Sandri

PREFACE III 序言三

CONTENTS 目录

Stylish & Fashionable 潮流时尚

010　GRAND COLONIAL　宏伟的殖民地风格公寓

022　W17 APARTMENT　W17公寓

028　SAN FRANCISCO SHOWCASE HOUSE　旧金山展示房屋

032　MINIMALISTIC ANIMAL MAGETISM　简约风格的魅力

040　19TH STREET APARTMENT　19街公寓

Creative & Chic 创意别致

048　LA APARTMENT　LA公寓

056　TRIBECA REGENCY　特里贝克区摄政风格住宅

062　LAUREL　Laurel公寓

072　RANCH HOUSE　牧场式房屋

082　WEST HOLLYWOOD RESIDENCE　西好莱坞住宅

092　ANTONIO CARLOS RESIDENCE　Antonio Carlos住宅

098　COASTAL COLLECTED　海岸收集之家

110　46 NORTH AVENUE,LEICHHARDT　46号北大街，Leichhardt

Comfortable & Cosy 舒适居家

120　LAKE SHORE PENTHOUSE　湖滨顶层公寓

128　DEPARTAMENTO POLANCO 1　Polanco 1公寓

136　CÁUCASO HOUSE　高加索住宅

144　UPPER WEST SIDE WATERFRONT APARTMENT　上西区海滨公寓

156　ONE HYDE PARK　海德公园一号公寓

164　DUNEIRER RESIDENCE　Duneirer 住宅

168　VIDALTA　Vidalta 公寓

176　CANARY WHARF　Canary Wharf 公寓

182　NEW ENGLAND STYLE　新英格兰风格住宅

Characteristic & Mixmatch 特色混搭

194　SOHO LOFT　Soho Loft 建筑

200　CLASSIC DESIGN　洛杉矶经典公寓

208　VENICE BUNGALOW　威尼斯小屋

214　WILTON RESIDENCE　Wilton 住宅改建

222　TERRACE 2　二号平台屋顶式建筑

228　HOLLYWOOD HILLS RESIDENCE　好莱坞山住宅

238　PRIVATE HOUSE IN MOSCOW AREA　莫斯科地区私宅

Modern & Simple 现代简约

248　URBAN FOREST　城市森林

256　GUNESLI PARK GARDENYA　Gunesli Park Gardenya 公寓

266　FIVE BEDROOM DETACHED FAMILY HOME　五间卧室的独立住宅

272　CRITZ RESIDENCE　Critz 住宅

278　RL HOUSE　RL 住宅

286　SEA TOWERS APARTMENT　海塔公寓

Exotic & Charming 异国风情

296　SOUTHAMPTON MOROCCAN　南安普敦摩洛哥风格住宅

306　LUXURY AT EASE　安逸舒适中的奢华享受

314　THE FACILITATION　简易化公寓

322　THE AMALGAMATION　融合风格住宅

330　PARK SLOPE REVIVAL　Park Slope 住宅"复兴"

338　WINTER HOME IN VERO BEACH　Vero 海滩的冬日之家

348　KUKIO RESIDENCE　Kukio 住宅

358　INDEX　索引

潮流时尚

STYLISH & FASHIONABLE

LOCATION
Singapore District 10, Singapore

GRAND COLONIAL

宏伟的殖民地风格公寓

DESIGNER *Nikki Hunt*
DESIGNER COMPANY *Design Intervention i.d.*
AREA   *1,100 m²*
PHOTOGRAPHER   *Jo Ann Gamelo-Bernabe*

The home is newly built, special attention had to be paid to the architraving, windows and door design, adding paneling and ceiling details to recreate the style. The brief was to recreate the grandeur and romance of life in colonial Singapore, while reinterpreting it in a fresh and contemporary way. The client requested formal public space, with a glamorous feel for entertaining and relaxed private spaces with a sense of easy living and an emphasis on freshness, infused with optimism.

Master Bedroom

To capitalize on the abundance of light, the designers selected a fresh palette. The pearlized wall covering, reflected light in a way that paint never can giving a warmer finish than a cold hard painted wall. Shots of blue balanced the heat of pinks and oranges. The shaggy rug and sheepskin throw added texture to the scheme. The curtains were hung high above the windows on acrylic rods for a light fresh feel. Their verticality drew the eye balancing the relatively short windows.

Dining Room

The walls were wrapped in a bold black and white stripe, hung horizontally for a modern twist. The chairs were covered in a chartreuse fabric echoing the hints of lime that peppered the formal areas. The designers designed the octagonal dining table to complement the odd contours of the room; its shape was echoed in the pattern on fabric.

Family Room

The designers retained a colonial feel, through the proportions and interior architecture. However, the feeling was very different to the entertaining areas. The palette was fresh and neutral creating a sense of tranquility. The designers selected distressed finishes, in contrast to the polished finishes in the formal rooms. The sofas are in distressed leather. Spills just wipe off and scuffs added to the patina. The coffee table was an old tree root.

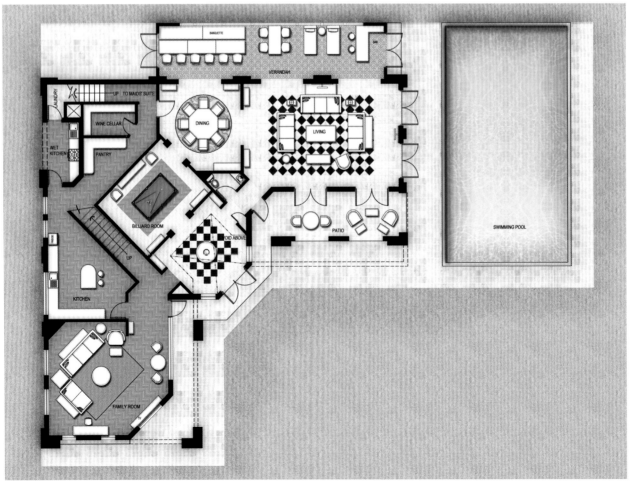

1st Floor Plan

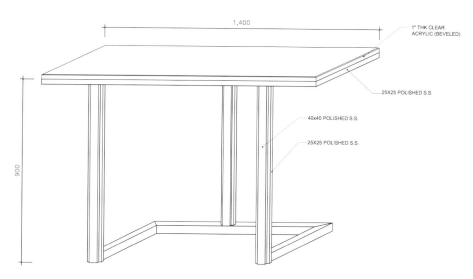

对于这所新建的住宅，门窗的头线条板和门窗的整体设计都需要格外的注意，并且要注意增加镶板和天花板的细节，以重塑空间风格。该设计的目的是要重现殖民地时期新加坡的宏伟壮丽和浪漫气息，并用一种清新感和现代感十足的风格对其加以重新诠释。住户要求室内有一个较正式的魅力十足的公共空间，作为娱乐场所。同时还要有一个可供放松和休息，充满生活气息，风格清新，传递积极、乐观的生活态度的私密空间。

主卧室

为了充分利用室内充足的照明，主卧室选择了色彩清新的配色方案。墙壁选择了珠光色的暖色壁布，以突出灯光在其表面柔和的反射效果，这是质地坚硬的涂料墙壁所不能企及的。冷暖色调的搭配使用，冷色系的蓝色色调平衡了暖色调的粉色和橙色。毛茸茸的地毯和羊皮毯子提升了空间的质感和品位。室内的窗帘高高地挂在腈纶材质的线绳上，给人以轻盈的感觉。笔直的垂直感不仅引人注目，也在一定程度上协调了长度相对短一些的窗子。

餐厅

餐厅的墙壁使用了横向的黑白条纹的壁纸，使装饰风格转向现代。椅子使用了黄绿色面料，与娱乐空间里柠檬绿的色调相呼应。餐厅选用了八角餐桌，进一步体现了独特的轮廓设计，也和椅子面料上的图案相呼应。

家庭室

通过室内设计和比例结构的调整，设计师保留了空间中的前殖民地风格，只是家庭室内的风格与娱乐区大相径庭。清新、中性的配色方案，营造出一种安静的氛围。仿古涂饰和娱乐区中经过抛光处理的涂饰形成对比。沙发面料为仿旧皮革。上面的油污已经被擦洗干净，磨损的表面加深了其古翠色的色泽。咖啡桌则是由一棵老树的树根改造而成的。

2nd Floor Plan

3rd Floor Plan

LOCATION
Chelsea, New York, USA

W17 APARTMENT

DESIGNER Karim Rashid
DESIGN COMPANY Karim Rashid Inc.
AREA 167 m²
PHOTOGRAPHER Jean-Francois Jaussaud

W17 公寓

The designer Karim has created a very hard Cartesian white "blank" gallery like space with white rubber epoxy floors, and a hint of fluorescent orange (The bathroom is fluorescent lime) for the front space and a pink carpet field for the back half. It seems that he is forever changing the space. Karim has been brought up with his father changing and moving around the furniture, paintings, etc. every month and he find he has the same habit. Generally, the furniture changes every month with new prototypes, old ones, production pieces, like a revolving on going dynamic gallery. The living space has Wolf Gordon wallpaper and new rug prototypes on a high gloss self-leveling epoxy floor. The kitchen is both minimal and maximal with stainless counter tops, clean lines but bright Abet Laminati plastic laminate cupboards. The office/lounge area has wall-to-wall broadloom carpet in light pink. Yellow glass walls separate the bathroom and the living space. The bathroom has a black high-gloss morphscape pattern laminate floor.

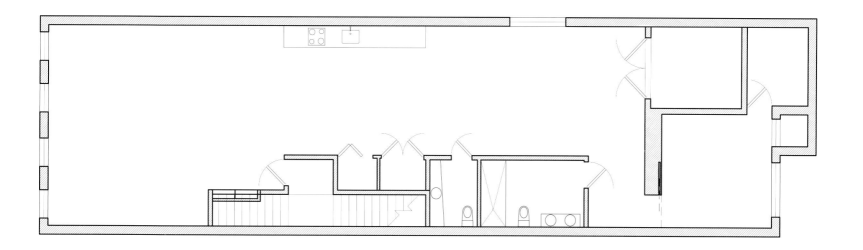

025

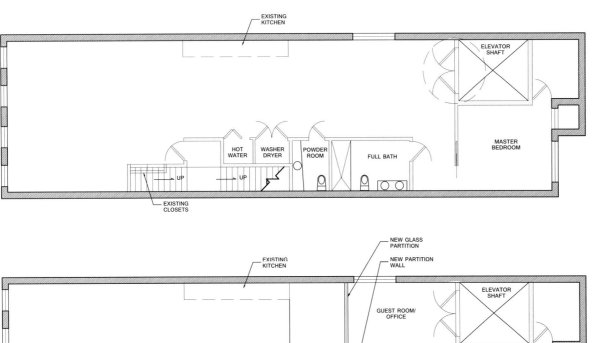

设计师 Karim 致力于打造一个笛卡尔风格的"白板"画廊般的空间。室内空间选用环氧橡胶材质的白色地面，并在前半部分的空间内加入了少量的荧光橙色元素。（浴室内选择了荧光青柠色），空间的后半部分铺设了粉色地毯。设计师似乎永远在改变着整个空间。设计师从小和父亲一起生活，父亲每个月都要改变家中家具和油画等物品的位置和布局，后来他发现这已经成为他的一个习惯。总体来说，每个月，家具，包括新的物品和原有的物品以及工业样品，都要进行位置调整，如同一个正处于循环更新中的动态画廊。生活空间内的墙壁选择了 Wolf Gordon 壁纸，自流平高光环氧树脂涂层地板上铺设了新式地毯。厨房中摆放着不锈钢案台，色彩鲜艳的 Abet Laminati 流线型层压塑料材质碗柜，既体现了最简单的装饰艺术风格，同时又彰显出室内空间整体装饰的全面性。办公区和休息区铺满了淡粉色的宽幅地毯。黄色的玻璃隔断分隔了生活空间和浴室，而浴室内选择了黑色变体图案的高光层板地面。

LOCATION
San Francisco, USA

SAN FRANCISCO SHOWCASE HOUSE

旧金山展示房屋

DESIGNER *Vernon Applegate, Gioi Tran*
DESIGN COMPANY   *Applegate Tran Interiors*
AREA   *111 m²*
PHOTOGRAPHER   *David Livingston*

The design inspiration for the Family/Media Room is to transform it into a dynamic and exciting space for entertainment and gathering. The strong and bold elements chosen excite the senses with an edgy yet elegant 21st-century style. Functionality is achieved by allocating space for various activities, while at the same time maintaining congruence throughout the different areas of the room. The atmosphere remains hip and modern andadvocates hassle-free lifestyle.

住宅中家庭室（媒体室）的设计灵感是要把它转化成一个用于娱乐和聚会的活力十足且令人兴奋的娱乐空间。在整个设计过程中，大胆和鲜明的设计元素唤起了观赏者对棱角分明且十分优雅的 21 世纪室内设计风格的整体感知。不同的空间具有不同的功能，各个空间在整体上又保持统一的风格。空间中洋溢着时尚、现代的气息，倡导一种毫无烦恼的生活方式。

LOCATION
Miami, Florida, USA

简约风格的魅力

MINIMALISTIC ANIMAL MAGETISM

DESIGNER *Edward Nieto*
DESIGN COMPANY Nieto Design Group
AREA 353 m²
PHOTOGRAPHER Ken Hayden

The living room features black glass slab floor, venetian plaster ceiling, modular art panel wall, 103" Runco TV, Kreon lighting, custom painted glass panels to cover column, custom circle sofa, custom desk in high gloss white lacquer, polished stainless steel trim, leather desk top, a pair of custom leather upholstered Fjord chairs, custom upholstered Fjord ottomans, custom patch rug in hair-on-hide & embossed cow-hide leather.

The master bedroom features black glass slab floor, venetian plaster ceiling, Kreon recessed lighting, modular art wall panel, white crocodile embossed leather Fendi king size bed, waterfall chandelier lamps, custom free-standing mirror with embossed crocodile leather frame, Alessandra chair custom upholstered in Spinneybeck black & white leather, Drift bench in white gloss polyurethane finish. The son's bedroom features a glass slab floor, Kreon lighting, venetian plaster ceiling, modular art panels, custom hair-on-hide wall panels, custom queen size bed upholstered in black crocodile embossed leather with polished stainless steel base.

The master bathroom features bathtub with zebra print hair-on-hide leather skirting, glass slab floors, Venetian plaster ceiling, Kreon lighting, glass wall panels inlaid with stainless steel, 3 rotating mirror storage cabinets, custom vanity mirrors with built in lighting, vanity & sink with slanted overflow to hidden drain, custom built frosted glass shower pan mounted flush with bath platform and floating drain system, Gessi fixtures & accessories. The powder room features black glass slab floor, black glass walls (some solid, some inlaid with stainless steel), custom mirror with back lit Plexiglas frame, recessed Kreon lighting, vanity with tap shelf faucet, zebra & stainless steel vanity stool, custom built-in cabinets with high gloss white lacquer touch latch doors.

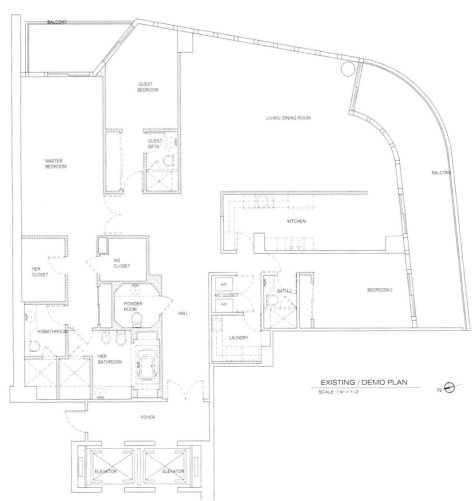

EXISTING / DEMO PLAN
SCALE: 1/4" = 1'-0"

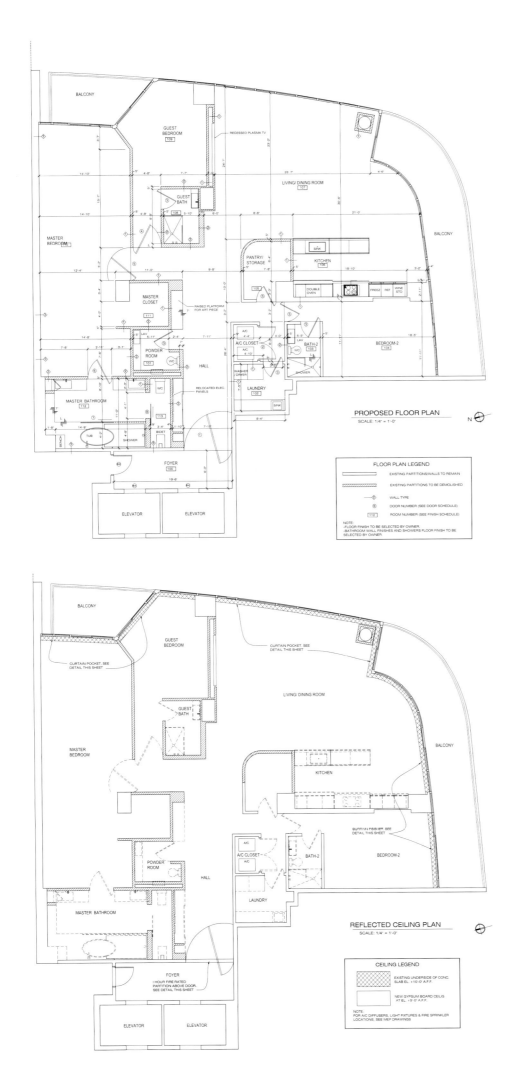

起居室选择了黑色玻璃地板、威尼斯风格的石膏天花板和模块艺术墙板。装饰元素包括：103英寸 Runco 电视，Kreon 照明，用于覆盖室内支柱的定制漆面玻璃嵌板，环形定制沙发，白色高光漆面定制书桌，经过抛光处理的不锈钢门窗镶边，皮质桌面，一对有皮质软垫的定制 Fjord 座椅，定制 Fjord 软垫搁脚凳，以及压花牛皮材质且皮毛一体的定制块状地毯。

主卧室也铺设了黑色玻璃地板，顶部是威尼斯风格的石膏天花板，墙壁选用模块艺术墙板。装饰元素包括：Kreon 嵌入式吸顶灯，Fendi 白色鳄鱼皮纹压花皮质特大号床，瀑布水晶吊灯，镜框为压花白色鳄鱼皮材质的自立式镜子，椅垫为 Spinneybeck 皮质的黑白色 Alessandra 椅，白色高光聚氨酯漆面漂流长椅。男孩房铺设了黑色玻璃地板，使用 Kreon 照明，顶部是威尼斯风格的石膏天花板，墙壁选用模块艺术墙板和皮毛一体的定制皮质墙板。空间中摆放着黑色鳄鱼皮纹压花皮质面料定制大号双人床，床的底座为抛光不锈钢材质。

主卫生间的设计亮点包括：浴缸旁放置的斑马纹皮毛一体的脚踏垫，玻璃地板，顶部的威尼斯风格的石膏天花板，Kreon 照明，玻璃镶嵌不锈钢材质墙板，三个带有旋转镜的储物柜，内置照明灯的定制化妆镜，与水槽一体的浴室柜倾斜式出水设计，以便隐藏排水管。定制磨砂玻璃淋浴室中的盘式的喷头与洗浴台、浮动的排水装置和 Gessi 固定装置和配件平齐。化妆间铺设了黑色玻璃地板，使用黑色玻璃墙（部分墙体质地坚硬，部分镶嵌有不锈钢），并放置透光定制树脂玻璃框架镜子，Kreon 嵌入式吸顶灯。浴室柜配有架式水龙头和斑马纹的不锈钢梳妆凳，内置橱柜上安装了白色高光漆面触点门闩。

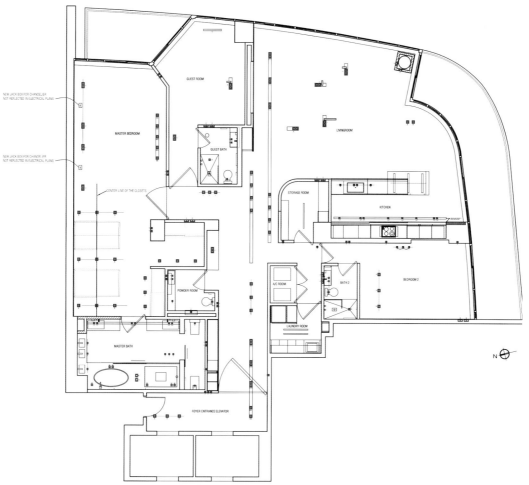

LOCATION
Chelsea, NY, USA

19街公寓

19TH STREET APARTMENT

DESIGNER Karim Rashid
DESIGN COMPANY Karim Rashid Inc.
AREA 149 m²
PHOTOGRAPHER Natan Dvir

Calm, Clarity, Cleanliness, Control. Ideally our domestic environments should exude positive energy, heightened experiences, contemporary design, a new comfort that is an extension of the new age of casualism, and spiritual well being—a place to enjoy, relax, socialize, work, and engage memorable experiences.

The spaces could be soft, curved, organic and more conceptual. Our surroundings should engage technology, visuals, textures, lots of color, as well as meet all the needs that are intrinsic to living a simpler less cluttered but more sensual envelopment. Rounded rooms are soft to provoke a more human friendly environment. Ideally, white is the overriding canvas for lightness of well-being and the neutral spaces that evokes calmness, casualness, with accents of vibrant glowing colors that are refreshing, awakening, and elevating, which are metaphors for a new global dynamism. Light plays a major role at changing the mood and hues of interior spaces.

Karim describes the apartment as one large space that is transformable, rearrangeable, and allows objects and furniture to breathe—to shape the personality of the space. The light-filled area is 149 m². Chelsea apartment has south and west facing windows with views to the Hudson River. Recently renovated, Karim replaced partition wall between bedroom and living room with translucent glass wall and sliding pocket door. Karim designed Marburg wallpaper covers the kitchen and living room walls. All other walls are painted white. Pink Lonseal Lonmetro UV flooring was laid in the living room, office and kitchen. Magenta carpet was paved in the bedroom and walkway in closet. The kitchen was renovated with custom white cabinets and Jenn-Air appliances.

宁静，通透，整洁，可控。理想的家居环境应该散发正能量，突出用户体验，展示当代的设计艺术，这种全新、舒适的体验是当代人追求随遇而安和幸福感的延伸。19街公寓就是这样一个可以享受、放松、社交、工作和体验美好生活的地方。

室内空间的设计柔和并含弧度，更好地体现了绿色环保和概念设计的理念。室内环境充分体现了现代科技在家居生活中的作用，丰富的色彩满足了居住者对视觉享受和高品质的追求。19街公寓能够满足人们想要脱离喧闹的群居生活、追求感官的苏醒和简单的生活环境的内在愿望。圆润的空间线条设计，打造出美好、和谐的居住环境。室内空间中的白色背景色好像帆布画的白色背景，令人倍感生活的舒适与轻盈。中性感十足的空间设计能够唤起人们内心的宁静和安逸，也凸显出室内其他醒目且令人振奋的鲜艳色彩。这样的室内配色方案也是全球化背景下的世界创新思维的体现。在整体环境中，灯光的设计对于空间内部氛围和色调的转换起着至关重要的作用。

身为设计师，Karim把这间公寓比喻成一个可转换、可随心安置的开阔空间。在这里，室内的所有物品都能够自由呼吸，正好突出了空间的个性化特征。室内采光面积可达149平方米。公寓内设有朝西和朝南的窗户，透过窗户可以尽情欣赏哈得孙河的风景。在这一次翻新设计中，Karim把卧室和起居室之间的隔墙替换成半透明的玻璃墙并安置了可滑动的内置门。厨房和起居室的墙壁则使用了由Karim设计的Marburg壁纸。公寓内的其他墙壁均涂成了白色。起居室、办公室和厨房则使用了粉色的Lonseal Lonmetro UV喷漆地板。卧室和步入式衣帽间中铺设了品红色地毯。重新设计后的厨房使用了定制的白色橱柜和Jenn-Air厨房用具。

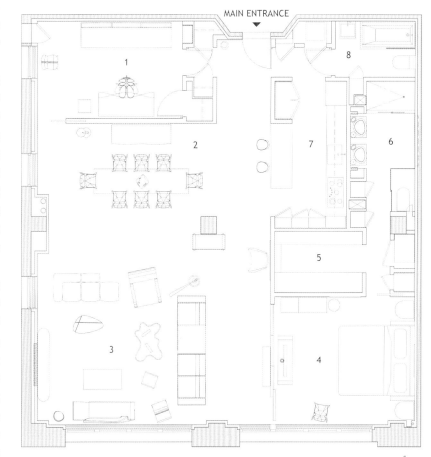

GENERAL PLAN
1 OFFICE
2 DINING
3 LIVING
4 BEDROOM
5 WALK-IN CLOSET
6 MAIN BATHROOM
7 KITCHEN
8 GUEST BATHROOM

创意别致
CREATIVE & CHIC

LOCATION
Belvedere, Belo Horizonte, Brazil

LA APARTMENT

DESIGNER *David Guerra*

DESIGN COMPANY David Guerra Architecture and Interior Design
TEAM Nínive Resende, Laís Machado
AREA 350 m²
PHOTOGRAPHER Jomar Bragança

When their first child started going to school, the couple bought an apartment in the city, leaving the country house, where they formerly lived, as a weekend destination.

The new home combines the cozy aspect of a country house and the urban and practical style of the big city. To attend the needs of the couple with two children, a renovation has been needed. The walls that divided the living room from the balcony were demolished, making spaces larger, more fluid and comfortable. The balcony became a gourmet bar/kitchen that can be used for the wine with friends and breakfast in family, with a view to the mountains. Linen sofa and chairs and a vintage armchair create a relaxing living area also in the balcony. A small fireplace has become a major element of the living room wall. The new warming ambience mix colors, rustic and natural materials with modern and technological ones. They are wool, natural linen, nude tones, leather in different colors—honey, whisky and chocolate—wood and demolition wood, gray concrete, Silestone rock, stainless steel, yellow metal, bronze, mirror, glass and acrylic, all materials that combined, create a great ambience.

The choices of the furniture noted the concern of creating a place that prioritizes comfort, warmth, elegance and relaxation. That way we can see a mix pieces both by Brazilian designers, such as Sergio Rodrigues, Pedro Useche, FredericoCruze, as by international designers, like De Padova, Minotti of B&Bitalia, Maxalto, Muuto and Mooi. The entire floor of the apartment, except the wet areas, has been replaced by wide planks of mahogany field brought from a farm. The floor has gone through a bleaching process, maintaining the identity and rusticity from the wood and giving a more light and modern touch to the place. On the wall, gray concrete and panels of different types of wood, such as mahogany field, pink mahogany, cedar and cinnamon, bring color and warmth to the room.

当这对夫妇的第一个孩子开始上学时,他们便在城市里买了这一套公寓,而他们以前居住的郊区的房屋只用于周末度假。

新家的设计风格结合了温馨的乡村住宅风格和大都市的实用装饰风格。为了满足这对夫妇和两个孩子的需求,整套公寓需要进行改装设计。设计团队首先拆除了阳台和起居室之间的墙壁,将两个空间进行了合并,增加了扩大后空间整体的流通性和舒适感。公寓中的阳台被改造成一个品尝室,可以在这里吃早餐,也可以和朋友饮酒、聊天,与此同时还可以欣赏室外的山色。亚麻布料的沙发和椅子以及老式的扶手椅,使改造后的阳台俨然成为一个可供放松的小客厅。起居室的小壁炉是其墙壁的设计亮点。在新设计中,选用了暖色调的混合配色方案,同时结合使用了粗糙、天然的装饰材料和现代高科技的装饰材料,包括羊毛,天然亚麻,裸色涂饰,蜜黄色、威士忌色和巧克力色的皮革,木料和可拆分的拼装木材,灰色混凝土,赛丽石英石,不锈钢,黄铜,青铜,镜子,玻璃和睛纶,所有这些材料的搭配使用,营造出一种宜人的情调。

家具选择方面考虑的因素依次为:舒适、温暖、优雅和放松。整间公寓的设计风格是巴西风格(如 Sérgio Rodrigues, Pedro Useche, Frederico Cruze 等著名巴西设计大师的装饰风格)和国际风格(如 De Padova, Minotti of B&Bitalia, Maxalto, Muuto and Mooi 等设计大师的风格)的结合。除了洗漱间等湿区之外,地板被全部换成从林场中买来的红盾籽木材质的宽木板。经过漂白处理的地板,既保留了地板的木质感和田园气息,又提升了整个空间的轻盈感和现代感。墙壁选用的灰色混凝土给室内带来了和谐之美。另外,室内使用的不同材质的木料,如红盾籽木、粉盾籽木、杉木和肉桂树料等,既丰富了空间色彩,又营造出温馨的氛围。

LAYOUT

LEGENDA:
01 - HALL
02 - LIVING ROOM
03 - DINING ROOM
04 - BALCONY
05 - GOURMET AREA/ BAR
06 - KITCHEN
07 - SUITE 01
08 - SUITE 02
09 - OFFICE
10 - MASTER SUITE

LOCATION
Tribeca, Manhattan, USA

特里贝克区摄政风格住宅

TRIBECA REGENCY

INTERIOR DESIGN Erin Fearins, Ward Welch, Catherine Brophy

ARCHITECT   Ward Welch
FIRM   CWB Architects
TEAM   Janelle Gunther, Sarah Ramsey
AREA   371 m²
PHOTOGRAPHER   Hulya Kolabas

An expanding family prompted the renovation of this 371 m², three-bedroom TriBeCa condominium. The primary objectives were to renovate and better define the kitchen, improve the flow between the kitchen, living, and dining rooms, and refine the entry. In addition, a powder room was expanded into a full bath. The scope of work also included interior design, rooted in a mid-century Regency aesthetic with distinct Asian influences and a bold color palette. Pre-existing arches were preserved, and built-in shelving was added to bring down the scale of the bedroom and establish a level of intimacy. It was important that the children's quarters both maximized layout and provided kid-friendly, built-in storage.

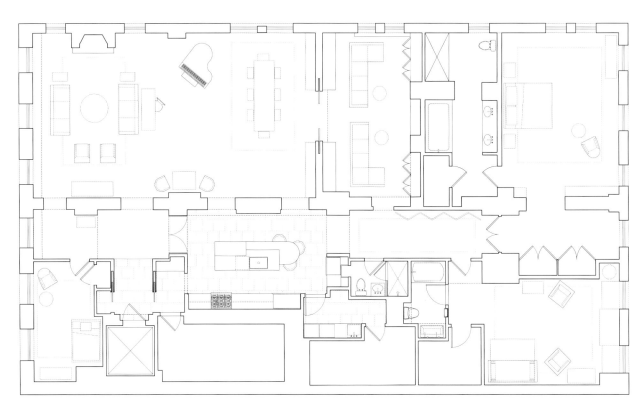

一个不断扩大的家庭推进了对这个占地 371 平方米并配有三间卧室的位于特里贝克地区的独立产权公寓的整改。该项目最初的目标是修整并更合理地界定厨房区域，改善厨房、起居室和餐厅之间的空间流线，并改善住宅的入口。另外，化妆室被扩大成一个包括厕所和沐浴区的完整的卫生间。

该项目也包括住宅的室内设计，其设计风格以中世纪摄政时期的审美标准为基础，带有鲜明的亚洲风格的印记并选择了色彩鲜明的配色方案。卧室内原有的拱门被保留下来，同时增加了内置搁架，在视觉上减小了卧室的面积，以增加卧室给人的亲近感。儿童区既最大限度地进行了装饰布置，又为儿童提供了友好型的内置储藏空间。

In this project, in order to achieve the characteristics needed by the new owners, there had to be a complete renovation of the house, developed in a single story, with the yard as main character of the space.

When approaching the project, the idea was to generate areas which would be both harmonious and functional for the family. The philosophy of the MARIANGEL COGHLAN firm is that well-accomplished interior design helps people to live better, not only because it creates beautiful environments, but because it manages to profoundly understand each person's way of life, helping them to resolve each area within the house in the best way. The light, both natural and artificial, in each part of the house was especially taken care of. The use of additional lamps to create light accents was an important part of the design concept for the project.

Each furniture piece was exclusively designed with the spaces in mind, carefully hand-crafted with the quality of Mexican craftsmen, in domestic woods. The upholstery was also made especially for the house, using materials from different places of origin. Decorative objects, mats, paintings, plants and other accessories were a proposal made by the firm in order to integrate the whole concept. "The most important thing when MARIANGEL COGHLAN design is to exceed their clients' expectations, creating spaces for encounter between persons."

LOCATION
Mexico City, Mexico

Laurel 公寓

LAUREL

DESIGNER *Mariangel Coghlan*

DESIGN COMPANY MARIANGEL COGHLAN
AREA 720 m^2
PHOTOGRAPHER Héctor Velasco Facio

为了彰显房屋新主人所要求的房屋特色，整个住宅需要进行全面的重新装修。整改在这个单层建筑中展开，并把房屋的院子作为整体装饰中的重头戏。

在整个项目刚开始运作时，设计师的目标是要为整个家庭打造一个和谐且功能强大的空间。MARIANGEL COGHLAN 公司的设计哲学是：好的室内设计能够为居住者提供更优质的生活，这不仅是因为它创造出了美丽的居住环境，更因为它深刻地了解每个居住者的生活方式，并帮助他们以最好的方式解决在居所中的每一个空间内所面临的问题。每个区域内的照明，包括自然光照和人工照明，都经过特殊处理。为了突出照明的效果，室内摆放了额外灯饰，这也是该项目设计理念中的一个重要组成部分。

室内空间中的每一件家具都是为了配合整体空间而特别设计的，并使用了国产木料，由墨西哥的工匠精心地手工制作而成。装潢品也都是特别定制的，并选用了来自不同产地的材料。装饰性的物品，垫子、油画、植被和其他的装饰物也是在设计公司的建议下选择的，以便与整个设计理念相融合。在 MARIANGEL COGHLAN 公司的设计中，最为重要的事情便是超出客户对室内设计的预期，创造一个人们能够在此邂逅的空间。

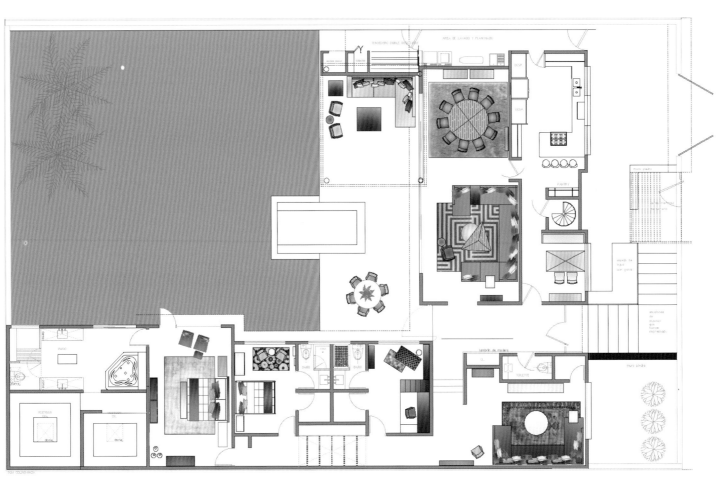

LOCATION
São Paulo, Brazil

RANCH HOUSE

DESIGNER Pablo Galeazzo

DESIGN COMPANY *Galeazzo Design*
AREA *1,900 m²*
PHOTOGRAPHER *Maira Acayaba*

牧场式房屋

Located in the countryside, near to the city of São Paulo, this ranch house of 1,900 m² was designed to be a place of leisure and meeting to several generations of a big family.

Natural and sustainable materials were chosen, such as demolition wood to the floor and cabinetry parts, bamboo for the roof overlay and pergolas, rough stone for the fireplace and some walls and soil and calcareous paint. In the main hall, an old log was brushed and used as axis for a helical staircase made of corten steel.

The house was divided in sections:

1. Spa: game room, massage room, steam room, dry sauna, turkish bath, leisure room, gym, heated pool with current and spa for 30 people, both with retractable roof.

2. Internal Social Area: dining room, lunch room, social hall, pub, cellar and movie theatre.

3. External Social Area: Big pool with spa, hamick area and a restaurant up to 100 people with infrastructure for bar area, barbacue and pizza, and a little stage for performances.

4. Intimate Area: With 7 suites and annex for guests.

For the decoration, furniture and lamps with contemporary design, rugs made of straw yields and wool, and vintage pieces.

这座位于农村的牧场式平房距离圣保罗市很近。房屋面积为1900平方米,可供包含几代人的大家庭休闲和聚会使用。

室内空间选择了天然和可持续发展的建筑材料。例如,地板和橱柜选择了拆分的组装木块作为材料,屋顶盖和绿廊则是由竹子制成的,壁炉和一些墙壁使用了粗糙的石料,以及土和石灰的墙面涂料。大厅中摆放着一根清理干净的旧原木,作为考顿钢材质的螺旋式楼梯轴线。

整个房屋被分成了4个区域:

1. 温泉疗养区:包括娱乐室、按摩室、蒸汽浴室、干蒸室、土耳其浴室、休闲室、健身室;容量为30人的暖水池,也可用于水疗,并配有开合屋盖。

2. 内部社交区域:包括餐厅、午餐室、交谊厅、酒吧、酒窖和电影院。

3. 对外开放的社交区域:可供水疗的大游泳池,hamick区,容量为100人并配有吧台、烧烤和制作披萨的基础设施和器具的餐馆,还有一个用于表演的小型舞台。

4. 私人区域:包括7个套房和一栋供客人使用的附属楼房。

室内空间选用了现代设计风格的家具和灯具,以及稻草和羊毛材质地毯和其他复古风格的装饰品。

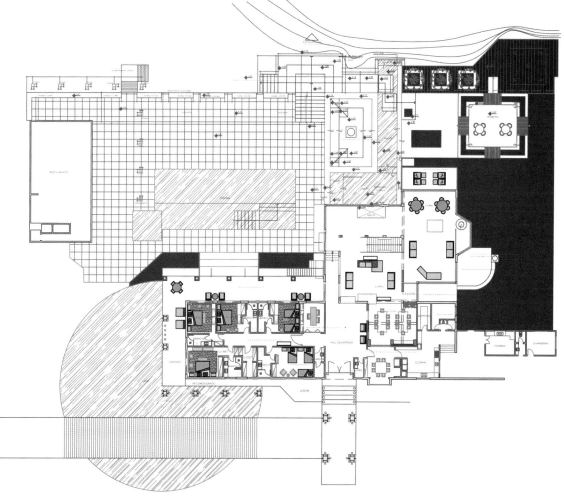

There is a point at which every designer finally gets to do for himself what he does for his clients. That point came for this designer two years ago. After 6 years of Hollywood Hills living, Tommy longed for his old "down in the flats" lifestyle. That's when he made the decision to move back into his first home in West Hollywood, a 1926 Spanish side by-side-duplex.

Tommy occupied both 116 m² units. The South unit was Tommy Chambers Interiors. The north unit was his home. With limited square footage it was imperative he rework most of the interior walls and spaces to make every square meter count. Tommy committed to an extensive 16-month renovation working with only the best trades and materials.

Design criteria: Tommy wanted to live with all his art, family heirlooms and books. Everything you see has a memory for him. How to make it all work together in the limited space was his biggest challenge. The layers made a very comfortable, coaster-free and cozy place to call home. Tommy feels that the arrangements and mixtures are modern in composition and flexibility. The space could accommodate 12 people for dinner, 6 adults for sleep or you could enjoy a quiet night in luxury.

LOCATION
West Hollywood, CA, USA

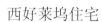

WEST HOLLYWOOD RESIDENCE

DESIGNER *Tommy Chambers*

DESIGN COMPANY Tommy Chambers Interiors, Inc.
AREA 116 m²
PHOTOGRAPHER David Phelps

在现实生活中有这么一点，那就是每一个设计师在给无数客户设计完房屋之后，总要给自己设计一套住房。两年前设计师 Tommy 就有了这个想法。在好莱坞山上居住 6 年之后，Tommy 渴望能够回归"平地"的生活方式。就是这个时候他决定搬回他的第一个家，位于西好莱坞的建于 1926 年的联排双层公寓中。

Tommy 的两个家的占地面积都为 116 平方米。位于南面的建筑作为 Tommy Chambers 室内设计公司，北面的建筑是他的住宅。由于建筑面积有限，Tommy 对室内大部分内墙和空间进行的改造是十分必要的，以充分利用每一平方米的面积。Tommy 耗时 16 个月完成房屋改造，聘请了最好的技工并使用了最好的建筑装饰材料。

设计准则：Tommy 想与他所有的艺术品、祖传遗物和书籍每天生活在一起。所以空间内的每一件物品，对他来说都有着特殊的意义。在整个设计中，Tommy 所面临的最大挑战就是如何在有限的空间内合理地摆放所有物品，使它成为一个远离海岸、温馨舒适的家园。Tommy 认为室内物品的布置和搭配组合体现了现代室内装饰艺术的灵活性。他的家中可以容纳 12 人就餐，供 6 个成人住宿，也可以独自在奢华的家中享受安静的夜晚。

The original layout of this apartment, two blocks away from the city's flagship avenue was originally subdivided into social, service and living areas. Looking for a more integrated layout for this mid-century apartment, architect and product designer Mauricio Arruda redesigned the spaces, using materials and finishing that were iconic of the 1950's aesthetics in Brazil and a unique mix of furniture. With an area of 150 m², the 60-year-old apartment revealed its potential to upgrade to the new needs when all the walls were knocked down and just the central pillar in the social area remained still.

In the living area, bathed in natural light filtered through the canopy of trees planted at the time of construction of the building, there are several pillows with fabrics of the Brazilian brand FAAUNA on the MICASA couch and George Nelson lamp. The armchairs and coffee tables as many other furniture and objects are part of the private collection of the resident, collector of Brazilian mid century modern furniture, most of them bought in flea markets. The dining room table HOUSE OF CARDS, designed by Mauricio Arruda, blends perfectly with the green folding chairs AIR by Tom Dixon. On the wall, more artworks by Brazilian artists such as Marcos Chaves, Marepe, Hercules Barsotti and Ernesto Neto, and a framed Werner PANTON Orange cloth brought from Berlin and a childhood photo of the architect next to his brother.

In the TV room, a sliding door allows the isolation of the space for the convenience of any guest. The Florence Knoll sofa works perfectly as bed when needed. In this area, the photo Marcos Villasboas dialogues with the pictures of the great Brazilian popular illustrator J. Borges. The center table in rosewood is an original piece of Brazilian master furniture designer Sergio Rodrigues. The white-wired iron chair by unknown author is an original piece from the 1960's, a classic item for Brazilian balconies and swimming pool areas of the time.

LOCATION
São Paulo, Brazil

Antonio Carlos 住宅

ANTONIO CARLOS RESIDENCE

DESIGNER *Maurício Arruda*

DESIGN COMPANY *Maurício Arruda Arquitetos + Designers*
AREA   150 m²
PHOTOGRAPHER   *Fran Parente*

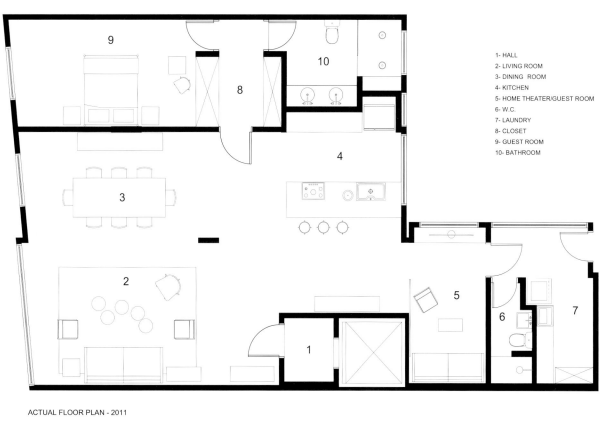

ACTUAL FLOOR PLAN - 2011

1- HALL
2- LIVING ROOM
3- DINING ROOM
4- KITCHEN
5- HOME THEATER/GUEST ROOM
6- W.C.
7- LAUNDRY
8- CLOSET
9- GUEST ROOM
10- BATHROOM

这座公寓距离市内的旗舰大道两个街区，最初被划分为社交、服务和起居区。为了给这个中世纪风格的公寓打造一个更加综合的布局设计，身为建筑设计师和产品设计师的 Mauricio Arruda，通过使用标志性的 20 世纪 50 年代巴西审美风格的材料和设备，并搭配独特的家具，对整个空间进行了重新设计。这座已有 60 年历史、面积为 150 平方米的公寓在拆除了室内的所有墙壁并且仅保留了社交区的中立柱之后，充分显示了其更新升级以适应新需要的潜力。

起居区沐浴在住宅建造之初种植的树木冠层滤过的自然光照下。MICASA 沙发上摆放着几个巴西 FAAUNA 品牌面料的抱枕，旁边是 George Nelson 灯饰。扶手椅、咖啡桌等一些物件都是爱好收集巴西中世纪现代家具的居住者的个人收藏品，它们中的大部分都是从跳蚤市场上买来的。餐厅内摆放着由

Mauricio Arruda 设计的 HOUSE OF CARDS 餐桌，完美地与 Tom Dixon 设计的绿色 AIR 折叠椅相匹配。墙壁上摆设了一些巴西艺术家 Marcos Chaves, Marepe, Hercules Barsotti 和 Ernesto Neto 等人的艺术作品和一幅从柏林买来的镶有外框的 Werner PANTON 橙色布画，另外还有一张设计师和他兄弟的童年合影。

电视间的拉门使客人能够享受与外界相隔离的空间。Florence Knoll 沙发在需要时正好可以当作床来使用。Marcos Villasboas 的照片与巴西十分受欢迎的插画作家 J. Borges 的相片形成对话关系。摆放在中央的红木材质的桌子是巴西著名家具设计大师 Sergio Rodrigues 设计的原作。白色的铁质金属线椅子是 20 世纪 60 年代巴西的家具真品，也是那个年代住宅阳台上和泳池边的经典配件，只是它的设计师已无从知晓了。

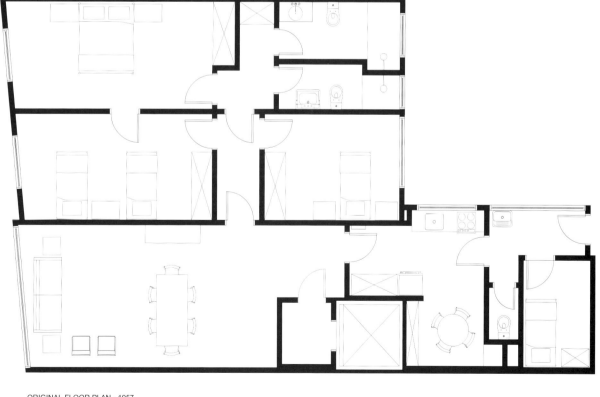

ORIGINAL FLOOR PLAN - 1957

LOCATION
Rancho Santa Fe, California, USA

海岸收集之家

COASTAL COLLECTED

DESIGNER *Kari Arendsen*
DESIGN COMPANY Intimate Living Interiors, LLC

Coastal Collected was inspired by its location Rancho Santa Fe, California. Located on one of the most beautiful coastlines in the world coupled with a sophisticated equestrian community, this ranch meets coast home represents this clients personal style perfectly. They are a young active family with many social ties in their community. With four children, Grandma, Mom, Dad, two dogs and a cat, this is a vibrant household. The design challenge was to create a highly functional space to accommodate their large active and very social family. The Dad is a professional baseball player and a pillar in their community, so he wanted a home that was inviting, comfortable and can accommodate as many guests as possible. The architect did just this, from the game room, to the dining room, and to the kitchen. The architect optimized the potential of this space to its fullest. By using built-ins that created multipurpose spaces, and using materials and finishes with refined yet resilient finishes, the architect were able to create a space to accommodate a highly functional and practical without sacrificing style (This is the most fundamental description of great interior design). All the textures, finishes, lighting and color palette were carefully chosen to highlight their personal style and worldly collectables, beautifully blending the ranch and coast lifestyle.

　　该项目的设计灵感来自于住宅的所在地——加州的兰乔圣菲小镇。位于世界上最优美的海岸线沿岸并置身于高端马术社区中，这个牧场和海岸风格相结合的住宅完美地彰显出居住者的个人风格。这是一个与社区社会关系联系紧密的年轻而活跃的家庭。这个充满活力的家庭包括四个孩子、祖母、妈妈、爸爸，以及两条狗和一只猫。该设计的挑战之处就是要打造一个高度功能化的空间，以适应这个善于社交的家庭的大量日常活动。家中的男主人是一名职业棒球运动员。作为整个社区的核心人物，他希望整个住宅能够洋溢着热情、好客的氛围，在提供舒适的家居环境的同时，能够尽可能多地接纳客人。从游戏房到餐厅，再到厨房，室内空间中每个角落的设计细节都是按照客户的要求去做的。在该设计中，设计师最大限度地开发了空间利用的潜能。与此同时，使用嵌入式结构来打造多用途的空间，并配有精致而有弹性的涂饰，既满足空间内高度功能化和实用性的要求，同时又不会牺牲住宅的风格。（这应该是优秀的室内设计最基本的类型）。所有的装饰结构、涂饰、灯光和配色方案都经过精心的挑选，以彰显业主的个人风格和空间内摆放的世俗珍藏品，并使牧场和海岸的生活方式完美地融为一体。

LOCATION
Inner West, Sydney, Australia 46号北大街，Leichhardt

46 NORTH AVENUE, LEICHHARDT

DESIGNER *Rolf Ockert*
DESIGN COMPANY Rolf Ockert Design
AREA 150 m^2
PHOTOGRAPHER Paul Gosney

The site is located in a heritage conservation area. Being only about 100 m² in size, the existing free-standing house was far too small for its intended use as a home for a young family with children. The only way to accommodate the intended brief was to build a two-storey addition, something that did not exist in the area.

The designer's approach was to maintain and preserve the existing house in its entirety while designing the addition to be ostentatiously modern. High-level slot windows to the side laneway allow tree and sky views whilst still maintaining privacy. These slots then became a major design feature that was then continued in the pattern on the side facade. The site was very narrow, meaning that the internal layout of the building had to be organized very simply. The linear stairs are not placed in line with the existing corridor of the old house but on the other side of the extension. As a result, the relatively small space reads to be quite generous.

To facilitate ease of construction whilst maintaining integrity of the original building, the existing house was kept completely intact and largely untouched. A glass side strip and roof lights were inserted between the old and the new, not only clearly defining the two but also bringing light into the core of the house. The project was special in so far as the client had a gallery for Australian aboriginal art. Not only did the design have to provide space suitable for the display of such art but the client was also very open and keen to have color as part of their new interior, something that many clients are very careful of.

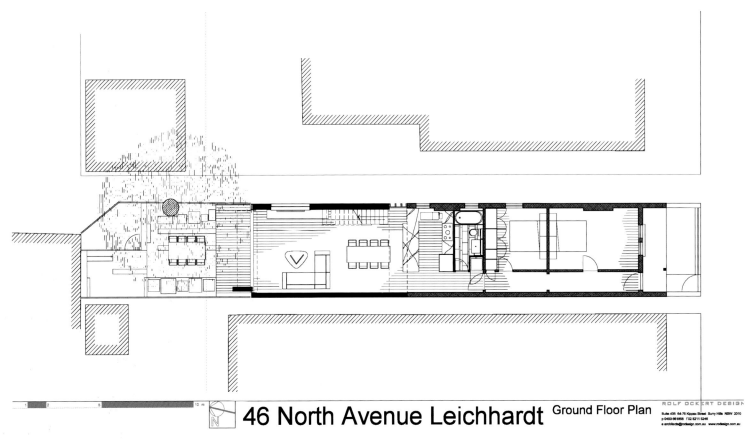

46 North Avenue Leichhardt — Ground Floor Plan

这座面积只有大约100平方米、位于遗产保护区内的独立式房屋,对于一对带有孩子的年轻夫妇来说实在是太小了。想要达到设计师预期要求的唯一办法就是扩展区域内原本不存在的空间,给房屋进行第二层的加层处理。

设计方案是从总体上完整地保留原有房屋,并将增加的第二层室内空间设计成炫目的现代风格。位置很高的槽式窗户和侧通道使室外的天空和树木清晰可见,同时又保留了空间的私密感。槽式设计成为该设计的重要特色,并被沿用在侧面的外观图案的设计中。由于室内的空间非常有限,这就意味着空间内的布局设计要简单。室内空间中的线形楼梯位于扩展区域的另一侧,不与原房屋中原有的走廊相一致,使相对狭小的空间显得大方、得体。

为了便于施工,同时保证原房屋的完整性,原有房屋的原貌被完好地保留了下来。在新旧建筑之间,增加了玻璃材质的隔离带和屋顶照明,不仅清晰地界定了新旧两部分建筑,同时也为房屋的中心区域提供了照明。该室内设计的独特之处是在室内空间中为居住者开辟出一个展示澳大利亚土著艺术的画廊,所以,该设计要为这样的艺术品的陈列提供适当的空间。与此同时,客户十分开放,与大多数住户在这一方面的谨慎考虑不同,他们非常热衷于在室内设计中融入丰富的色彩搭配。

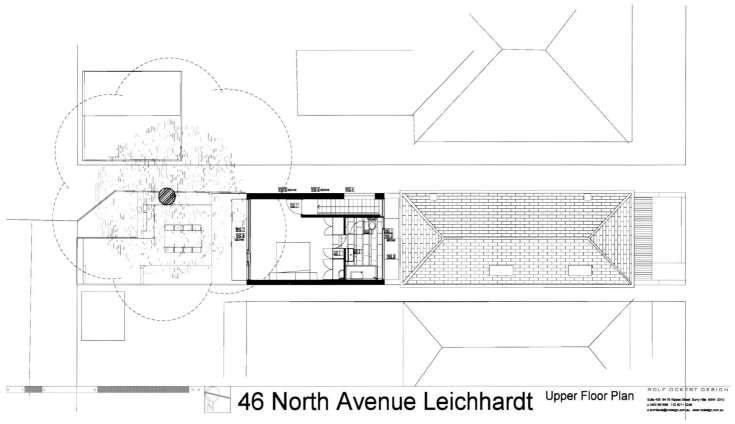

46 North Avenue Leichhardt Upper Floor Plan

舒适居家

COMFORTABLE
& COSY

LOCATION
Lake Shore Drive, Chicago, USA

LAKE SHORE PENTHOUSE

DESIGNER Jessica Lagrange

DESIGN COMPANY Jessica Lagrange Interiors
PHOTOGRAPHER Tony Soluri

湖滨顶层公寓

Jessica Lagrange Interiors was commissioned by a prominent Chicago lawyer and his wife to design their spectacular 836 m² unit in a twenty-six story residential building on Chicago's legendary Lake Shore Drive. The stunning building include a stone tower with a zinc mansard roof honoring French history. For this significant penthouse unit, which was used as the couple's city pied a terre, the designers took cues from the building's French Beaux Arts character that at first glance looked as transported directly from Boulevard Haussmann in Paris.

The general layout of the unit was based on the neoclassical principles of symmetryly relating to axes, the hierarchy of spaces, and harmonic proportions. The "Rotunda" became a significant connection between the public (gallery and entertainment) and private (guest suite and office) areas. A mural created by Scott Waterman titled "Ice Breaking" wraped this elliptical space.

The expansive Living Room, with its breathtaking Lake Michigan views, was adjacent to the peeky cypress paneled Library and the large round Dining Room housed in the corner "turret". The gold leafed eglomise gracing the walls and ceiling of the powder room brought a touch of glamour. Elegantly-detailed painted paneling, antiqued oak herringbone floors and custom hardware on double butternut doors were a few of the elements that contribute to the home's unique character.

The furnishings chosen were an exquisite mix of unique antique pieces, like the 19th-Century Venetian grotto console table, and new classic pieces such as the Victoria Hagan St. Simone chairs upholstered in hair-on zebra hide. Custom furniture and light fixtures, such as the gildes "Wisteria" chandelier and sconces featured in the Dining Room play an important role throughout.

123

Jessica Lagrange 室内设计公司受芝加哥一位知名的律师和他妻子的委托，对令人叹为观止的位于芝加哥的传奇湖畔快车道旁的 26 层住宅楼中面积为 836 平方米的公寓进行装饰设计。这座令人惊叹的建筑包括一座向法国历史致敬的、带有折线形镀锌屋顶的石制塔式建筑。这栋宏伟的顶层公寓单位，是这对夫妇在市内的临时住所。在对室内进行设计时，设计师借鉴了法国学院派的设计风格，使整栋公寓第一眼看上去如同是直接从巴黎的奥斯曼大道搬运过来的一样。

房屋单元的总体设计以古典主义的轴线对称美、空间等级和层次，以及和谐的面积比例为基础。房屋内的圆形大厅成为室内公共区域（画廊和娱乐空间）和私人区域（客人套房和办公区）的重要连接部位。一幅由 Scott Waterman 创作的名为"打破僵局"的壁画包围着这个椭圆空间。

可以欣赏密歇根湖美景的宽敞的起居室，与视力范围内的柏木建造的图书室和位于"塔楼"角落中宽敞的圆形餐厅相邻。墙壁上的金箔画屏为其增添了高雅的韵味，化妆室中的天花板也使整个空间变得更加迷人。灰胡桃木材质双扇门上的定制金属构件也是突出室内设计风格的设计元素之一。

设计师为房屋精心选择了独一无二的古董家具混合搭配。例如，19 世纪威尼斯风格的石制桌案，也包括全新的经典配饰，如装有皮毛一体的斑马皮草软垫的 St. Simone 座椅（由 Victoria Hagan 公司出品）。定制家具和照明设备，例如镀金的"紫藤"枝形吊灯和突出餐厅风格的壁灯，都在整体风格的塑造中起到了十分重要的作用。

LOCATION
Mexico City, Mexico

Polanco 1 公寓

DEPARTAMENTO POLANCO 1

DESIGNER *Claudia López Duplan*

DESIGN COMPANY López Duplan Arquitectos
AREA 370 m^2
PHOTOGRAPHER Héctor Armando Herrera

This apartment is located in Polanco, in Mexico City, just across from a park so the views are mostly of fantastic tree tops. The apartment has one level with wide spaces that allowed making a proper distribution of the areas resulting in a casual, very warmth and open ambience.

The remodeling project of this space considered a master bedroom for one person and a guest room; it also includes a library with enough privacy that, when necessary, may also be an additional guest room. Because of its location, the entrance hall lacks of natural light; this situation was solved with a plafond where lamps were installed and the walls were painted in an intense red color that unifies the walls and doors generating an ample single space. All the floors, except for kitchen and bathrooms, are of a light-colored wood, the same shade was maintained in all the custom made furniture. The plafond was placed as high as possible to avoid affecting the amplitude of the spaces.

The intensity and brightness of the entrance blends with the rest of the areas, for these, a neutral color was selected to make the most of the natural light. There are also some red accents in the same color of the entrance hall. The artificial light is indirect and where the height was not sufficient they were installed on furniture and walls. Some walls have accent lights to emphasize the space and there are also some suspended lights. Among the living room and dining room is a large column, which was covered with a luminous material, to convert it into a great central light object.

公寓坐落在墨西哥城的波朗科区，位于一个公园的对面，所以从室内向室外放眼望去，可以欣赏树顶的美丽景色。公寓只有一层，但室内空间宽敞，使设计师可以在这个整体空间内进行合理的区域分配，以打造一种开放、休闲且温馨感十足的空间氛围。

该项目的改造任务包括对一间单人使用的主卧室、一间客房和一间私密感十足的藏书室（如果需要的话，可以作为一间额外的客房）进行改造。由于地理位置的原因，门厅缺少自然光线的照射。这一问题通过在装饰性的天花板内安装内部照明灯而得到了解决。墙壁选择艳红色的涂饰，统一了墙壁和门的色调，使室内形成了一个宽敞的单一空间。除了厨房和浴室之外，其他空间内的地面都使用了淡色木料，相同的色系也被应用在了所有定制家具上。顶棚天花板被安置在尽可能高的高度，以免破坏空间整体的广阔感。

色彩浓烈、明亮的门厅与其他区域相融合，在这些区域中选择了中性色的配色方案，以充分利用自然光照，同时也在一些区域中使用了同门厅一样的艳红色的配色。人造光线则选择了间接照明，当室内的高度不足时，就将它们安装在墙壁或者家具中。一些墙壁上安装了强光灯以凸显整个空间，同时也安装了一些悬浮灯。在起居室和餐厅中使用了一个大的圆柱形照明灯饰，表面覆盖了夜光材料，将其转换成一个优质的中心照明物。

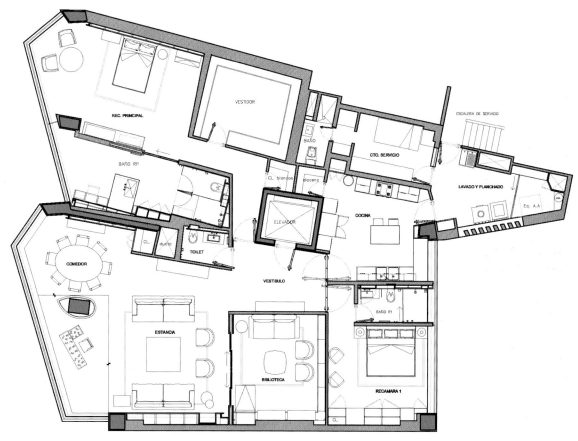

PLANTA

LOCATION
Mexico City, Mexico

CÁUCASO HOUSE

DESIGNER *Mariangel Coghlan*
DESIGN COMPANY MARIANGEL COGHLAN
AREA 790 m²
PHOTOGRAPHER Héctor Velasco Facio

This house, located on the west side of Mexico City, is a project that was undertaken for a young couple with two children. The goal was to create a home full of light, with large but cozy spaces, taking advantage of the natural light available and the views to the exterior of the house.

The selection of materials, textures and colors was the result of meetings between the owners of the house, the architect and the interior design team of the firm. The color scheme was chosen to give peace and tranquility to each of the areas of the house. An important aspect of the design was the selection and the placement of the furniture, rugs and most importantly the pieces of art. Every item of furniture used in the house was carefully designed by the firm; MARIANGEL COGHLAN and manufactured in Mexico. Nationally produced wood of varying types, was used with the intention of giving versatility to the spaces.

"A house is like a suit, the best ones are made to order". This is the philosophy employed in every project, working closely with the taste, needs and above all the dreams of every client. Every detail of every project is important, and the clients own possessions as well as the new designs, are carefully placed and integrated in order to achieve the concept.

这座位于墨西哥城西面的住宅，是为一对带着两个孩子的年轻夫妇打造的设计项目。该设计的目标是打造一个光线充足、宽敞又充满温馨感的室内空间，并充分利用可利用的自然光线和室外的景色。

室内空间的装饰材料、构造和色彩选择，是在房屋主人与建筑师和室内设计团队会面之后决定的。配色方案的目标是营造静谧、安逸的空间氛围。室内设计的另一个重要方面是室内家具、地毯和更为重要的艺术品的选择和摆放。公寓内使用的每一件家具都是由MARIANGEL COGHLAN设计公司精心设计并在墨西哥制造完成的。本国产出的各种类型的木材在设计中得以使用，以增加整个空间的多样性。

"住宅和套装一样，最好的都来自定制。" 这是建筑师和室内设计团队在设计每一个项目时的设计理念。因此，在每一次的设计中建筑师和室内设计团队都会密切关注每一位客户的品位、需求和梦想，并重视设计中的每一个细节。客户的私人物件和新陈设的设计作品在整体布局中都得到精心的安置并使它们相得益彰，以达到设计理念的要求。

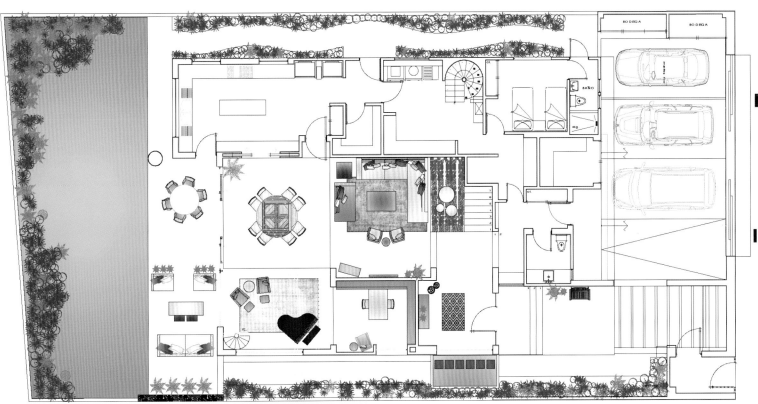

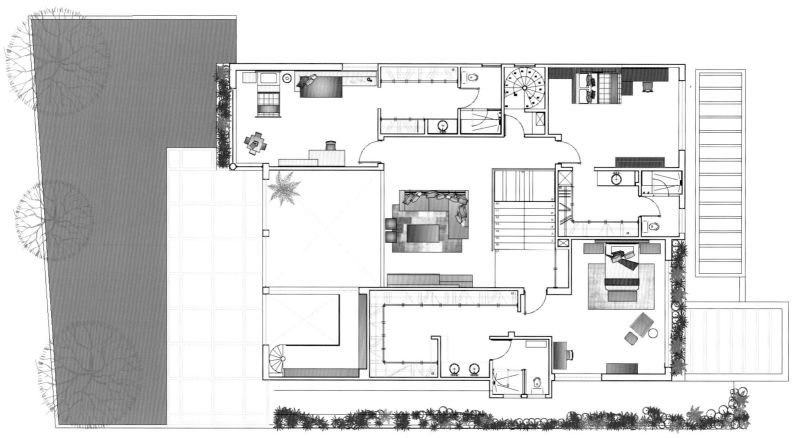

LOCATION
Manhattan, New York, USA

上西区海滨公寓

UPPER WEST SIDE WATERFRONT APARTMENT

DESIGNER Susana Simonpietri

DESIGN COMPANY   Chango & Co.
AREA   186 m^2
PHOTOGRAPHER   Jacob Snavely

Located on the waterfront of Riverside Boulevard in the Upper West Side of Manhattan, this newly-built 186 m² shell lacked all the charm of an older building. The family which was to inhabit it, however, were charming enough to make up for the blank canvas. So, the designers set off to bring a bit of each of them into their new home. Easy to work with to the extreme, their only requests were storage and a color palate suggested by the children for each their rooms. The storage request was answered in the form of 18 m of floor to ceiling custom built ins which help house toys, televisions and other equipment. The children rooms were an homage to childhood. The result of combining all these elements was a home which was easy to slip into at the end of the day, thoughtfully designed around the normal wear and tear of a young family, and just plain fun to play in.

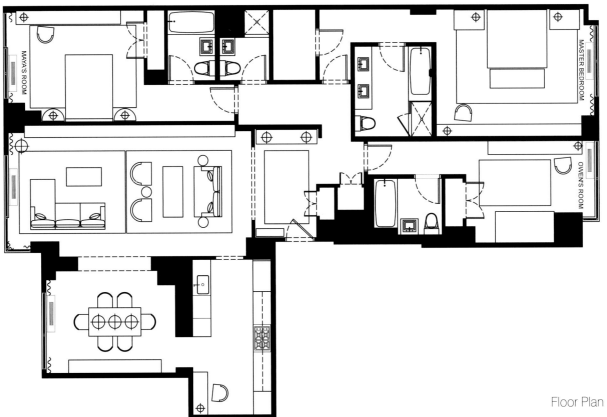

Floor Plan

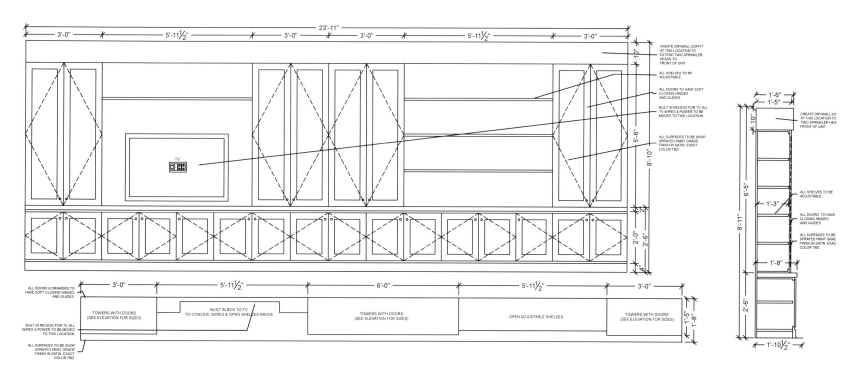

Living Room

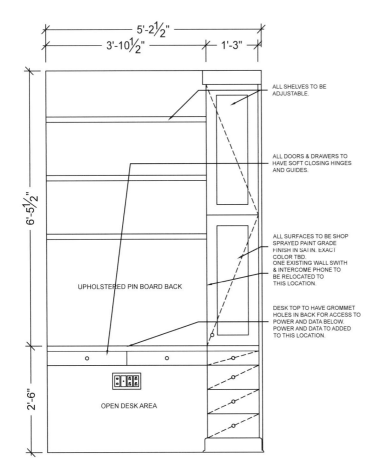

位于曼哈顿上西区江滨道的河畔,这座新建的占地面积186平方米的框架建筑不具有老房子所特有的魅力。而该公寓的新住户,却拥有十足的魅力去填补这一不足,所以该设计把每个家庭成员的个人元素融入其中。住户一家人的要求只有两点:一是公寓内要有足够的储物空间,二是根据孩子们的喜好,为他们的房间搭配他们喜爱的颜色。为了满足他们的储物需求,设计师安装了一个从地板到天花板18米长的定制嵌入式柜体,用以储存家中的玩具,容纳电视和其他设备。儿童房的设计是向童年的美好时光致敬。这些设计元素的组合搭配最终向住户提供了一个在外奔波了一整天后,可以轻松融入并得到全身心放松的温馨的家。通过细心的设计,该设计对这个年轻的家庭可能造成的日常磨损做了必要的防护处理,使他们在玩耍中能够尽情享受其中的欢乐。

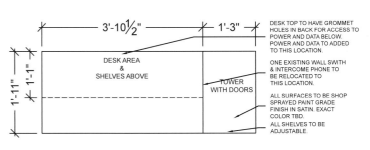

Kitchen   Office

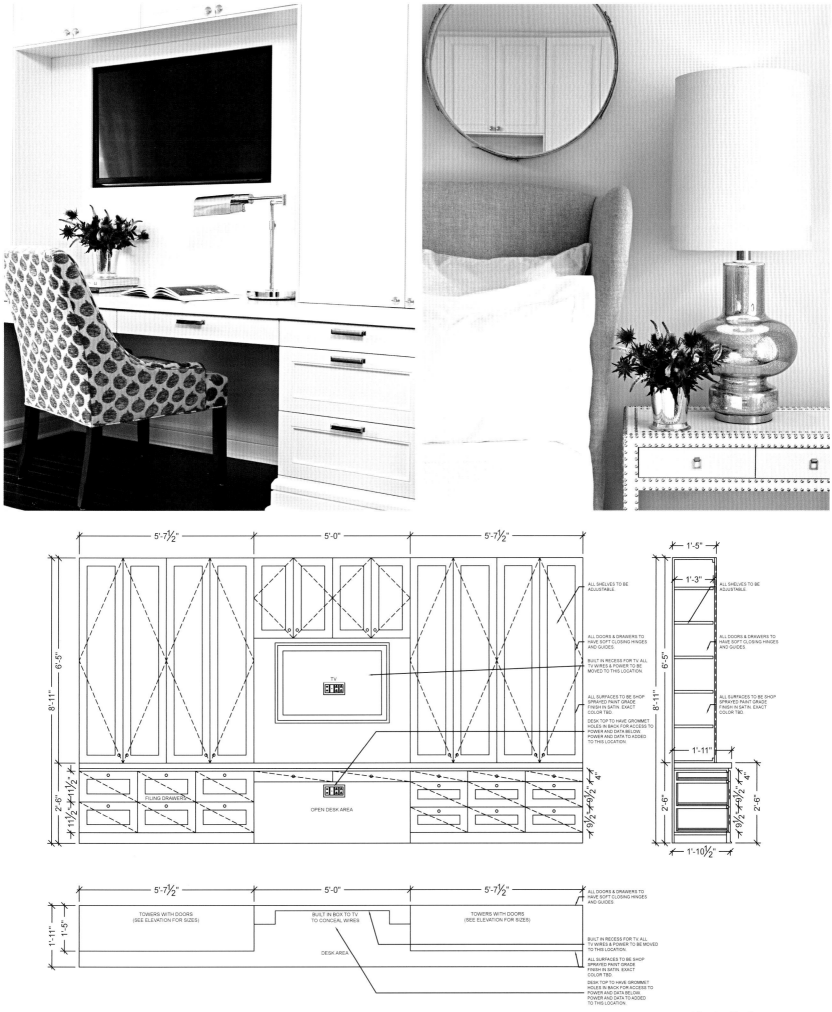

Master Bedroom

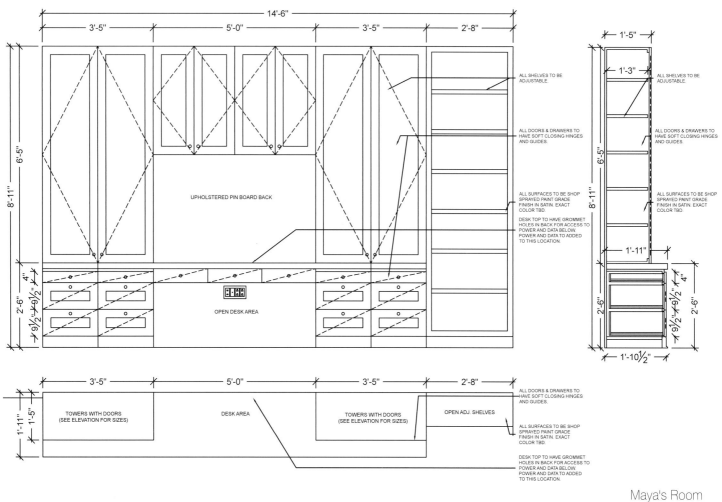

Maya's Room

LOCATION
One Hyde Park, London, UK

ONE HYDE PARK

DESIGNER *Casa Forma*

AREA 372 m²
PHOTOGRAPHER James Balston

海德公园一号公寓

When creating the interior design scheme for this 421 m² apartment in One Hyde Park, Casa Forma's focus was on practicality and functionality, whilst at the same time exuding the very best in bespoke luxury and elegance. The result was a sophisticated and timeless design, and a comfortable family home.

The immediate focus on entering the apartment wa the long corridor leading to the main living space, with its magnificent views of Hyde Park itself. Two features served to particularly distinguish the corridor. Firstly, high-gloss sycamore panels added light and subtlety. And then, on the other side of the panel, an abstract map of the Hyde Park pathways—the bronze clasps and metalized panels and resin of which combine to create a subtle 3D effect.

The furniture was made from darkened solid Indian Rosewood in piano high gloss finish, and with the bespoke leather sofa able to convert into a large guest double-bed, this room could also easily be turned into a bedroom. The introduction of a Swedish dry sauna in the en-suite bathroom linked to the study room was one of the most challenging aspects of this project. But thanks to Casa Forma's strong architectural acumen, this small yet complex feature turned out to be an extremely successful addition to the property. In the dining room, the bronze profile and glass joinery was framed by a subtly backlit tortoise shell and tiger eye mosaic, while exquisitely colored brown glass chandelier pendants hang over the dining table. The interior design of the main living space focused on maximizing those incredible views of Hyde Park. Accents of beveled glass relief on mirrors framing all the joinery added an Art-Deco touch to the room.

Through research, quality of materials, precision, and attention to detail all embodied perfectly in one property—making this Casa Forma project an outstanding example of considered, uncompromising and luxurious interior design.

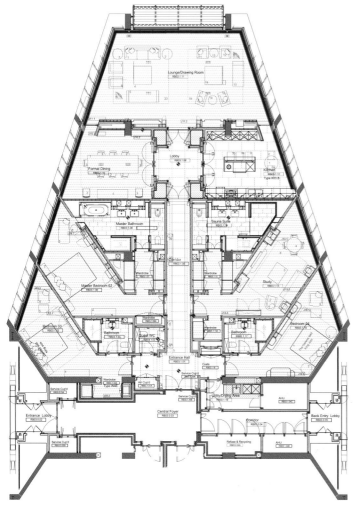

在对这个位于海德公园一号居住区、总面积为421平方米的公寓进行室内设计的时候，Casa Forma设计公司侧重于空间设计的实用性和功能性，同时使整栋公寓散发着特有的奢华感和优雅感，最终呈现给大家一个精美、经典的设计范例和一个舒适的家庭住宅。

进入公寓，首先映入眼帘的是直通主居住区的长走廊，置身其中可以欣赏海德公园优美的景色。走廊的设计亮点主要包括：选用的高光喷漆的美国梧桐木木板提升了整个空间的亮度和精妙感。木板的一端镶嵌着进入海德公园的通道缩略地图，地图上的铜钩和金属板以及其树脂层，共同创造出精妙的三维立体效果。

家具是由质地坚硬的深色印度紫檀木制成的，表面喷有钢琴表面使用的高光喷漆。定制皮质沙发可以折叠成一个双人床，整个房间也可以轻松地转变成一间卧室。该设计中最有挑战性的一项就是将瑞典干蒸室引入与书房相连的卧室中的独立卫生间中。归功于Casa Forma设计公司高超的建筑技艺，这个虽小但十分复杂的特色设计对住宅的整体设计具有锦上添花的作用。餐厅中，青铜材质家具和玻璃细木工制品上均安装了做工精细的透光龟壳材质和虎眼马赛克材质边框。餐桌上方悬挂着精致的褐色玻璃材质的水晶吊灯。主卧设计着重于使居住者尽可能多地欣赏海德公园内优美的风光。细木料制品上都镶嵌着雕有锥形玻璃浮雕的镜面边框，使整个空间彰显出Art-Deco建筑风格的特色。

该室内设计集深入的调查、高品质的装饰材料，精准的测量和细致的装饰细节为一体，使Casa Forma设计公司的设计作品成为一个被广泛认可的、高品质且奢华感十足的室内设计典范。

LOCATION
Manhattan, New York, USA

DUNEIRER RESIDENCE

DESIGNER *Pepe Calderin*
DESIGN COMPANY *Pepe Calderin Design*
AREA 279 m²
PHOTOGRAPHER *Barry Grossman*

The Duneiers wanted a home that was greatly modern, but still felt warm. By incorporating modern furnishings with warmer finishes, like dark woods, and natural stones, the designer was to able to create a space that maintained a balance between form and function. It was important to establish a kind of energy in the home that was reminiscent of the cosmopolitan city in which it is located in New York. Pattern, color, and light added the vibrancy it needed. However, while the majority of the apartment was designed to emulate the energy of the city, the Master Suite was designed to be just the opposite. A true retreat, the monochromatic white palette was given life through variety in texture, reflectivity and tonal patterns. Starting in the entry, Calderin skillfully mixed backlit onyx and stainless steel with sculptural, wave-designed wall of tobazzo-finished oak wood and LED lighting. The same tobacco-finished oak continued into the main living spaces, creating integral architectural features that forma look that was cohesive, elegant and warm. "It is a great impression", the designer says.

Duneier 一家希望他们的住所既有十足的现代感，又洋溢着家的温馨气息。通过现代装饰材料和暖色调涂饰的搭配使用，如深色木料和天然石材，设计师 Calderin 打造出一个形式和功能相协调的空间。坐落在纽约这样一个时尚大都市中，空间应该具有十足的活力感，与整个城市的气息相符。格局、色彩搭配和灯光都使整个空间活力四射。虽然空间的整体设计风格是大都市的鲜活和动感的真实写照，但主卧室的风格却迥然不同。如同一个真正的避风港湾，单一的白色色调通过使用不同的纹理以及各异的反射和声律，变得生动起来。从入口处开始，设计师就巧妙地搭配使用不锈钢，以透光玛瑙和淡烟草色的带有雕刻般波浪花纹的橡木和 LED 照明作为辅助。与此同时，大部分空间都延续使用了相同的淡烟草色橡木，以形成统一的装饰风格，整个空间既给人以统一感，又看起来优雅、温馨。"它给人留下极深的印象"，设计师本人这样评价，"它为整个空间营造出一种酒店式的氛围，也定义了整个空间的基调。"

LOCATION
México D.F., Mexico
VIDALTA

DESIGNER Elías Kababie
DESIGN COMPANY Kababie Arquitectos
AREA 150 m²
PHOTOGRAPHER Allen Vallejo

Vidalta 公寓

For the interior design of this apartment, it is generated a full atmosphere that describes perfectly the lifestyle of the family who lives it. One of the major challenges is the selection of furniture and accessories that, in combination with the finishes and the color palette, shaped each one of the different ambiances. The first thing that catches your eyes in the hall is a wall covered with copper foils that leads the view in to the apartment.

The living and dining room area is flanked by large windows on one side and the hallway leading to all areas of the apartment on the other. To increase light and height, a mirror plafond is installed across the whole ceiling resulting in a dramatic effect, intensified by the glare of the light marble floors. The dining area is emphasized with a long row suspended lamps with a slight variation in height to give a subtle movement on top of the rectangular and square tables that conform the dining table. In this space also stands out the combination of leather, wood and glass of the different furniture in both environments, creating an atmosphere full with dark and amber colors and textures broken only by the contrast of the warm purple carpet in the living room. The thick wall that divides the living room from the family room is covered with slabs of black stone whose motion gives a nice texture enhanced by the light from windows and ceiling.

One of the most important concepts for the design of the kitchen of this apartment is the incorporation of details that refer to the tradition of the family differ from the established parameters. On the ceiling is placed a classic Persian carpet framed in acrilyc, giving the room the warmth of the fabric and textures, and a bold splash of color. Another important issue is the distribution of the cooking areas in accordance with the canons of the religion, for which two separate islands are installed, one for the preparation of food that includes meat and other for the preparation of food including dairy products.

该设计完美地展现了居住在其中的一家人的生活方式。该设计面临的一大挑战是家具和装饰品的选择，并搭配配色方案和涂饰，在不同的空间内营造出不同的氛围。进入门厅后首先映入眼帘的便是覆盖着镀铜金属薄片的墙壁，一路引领你进入公寓。

起居室和餐厅的一侧墙壁上安装了大窗户，另一侧的走廊则通向各个房间。为了增加空间亮度和视觉高度，在顶棚上安装了装饰镜面天花板，十分引人注目，耀眼的浅色大理石地面也使其得到了进一步凸显。餐厅的设计亮点是悬挂的一排长长的挂灯。挂灯高低不同，悬挂在与餐厅风格相一致的矩形餐桌上方，尽显细微的变化。皮革、木材和玻璃材质家具，共同营造出一个以深色和琥珀色为主的整体空间，而只有起居室内的暖色调紫色地毯与其形成对比，打破了单一的色系。分隔起居室和家庭室的厚墙壁以黑色石板作为表面。在来自窗外和天花板的光线的映衬下，墙壁整体的质感得以提升。

厨房设计最重要的设计理念是室内装饰细节的合并，以体现居住家庭的传统。天花板上悬挂着一个亚克力镶框经典波斯装饰毯，带给空间一种温和的质感和一抹鲜艳的色彩。设计中的另一个重要问题便是根据业主所信仰的宗教教规来分配烹饪区，因此在空间内安装了两个独立的工作台，一个用于准备包括肉类在内的食物，另一个则用于准备包括乳制品在内的食物。

PLANTA ARQUITECTÓNICA

PROYECTO: VIDALTA
AGOSTO DEL 2011
PLANTA ARQUITECTÓNICA
ESCALA 1:125

175

LOCATION
Canary Wharf, London, UK

CANARY WHARF

DESIGNER *Casa Forma*

AREA 325 m²
PHOTOGRAPHER James Balston

Canary Wharf 公寓

Daring and sensuous Casa Forma's interior design for this 325 m² Canary Wharf apartment breaks barriers between living and entertainment. The exquisite residence transforms into a luxurious entertainment space at the flick of a switch. The element of fun is woven through the scheme by the use of unique materials, lighting, visuals and technology.

In the lobby, the visitor is greeted with spectacular floor to ceiling back-lit onyx panels framed in polished brass. The space features a Casa Forma console, 24k gold table lamps and a bespoke Italian marble floor with brass inlays. The main space of the apartment has been enlarged and sectioned into a dining area, kitchen and bar offering a breath-taking panorama of London. The dining area is backed by an impressive wall-to-wall joinery clad in metal tattooed glass, exclusively developed for this Casa Forma design. The bespoke glass and brass dining table offers the perfect setting for a fun dinner.

The kitchen area comprises an ample island with a fantastic bespoke chandelier along its entire length. The area in front of the island transforms into a dance floor at night, complete with illuminated glass floor. The bar area combines a range of extraordinary materials: back lit onyx, dark contrasting timber, perforated tarnished bronze and mirrors. The bar features a full blown LED display that is synchronised with the music creating a sleek private club atmosphere. The corridor in the left wing of the apartment is entirely clad in handmade brass-leafed mirror tiles creating a maze leading to the pool room. The Pool Room is a fun space with water bubble walls and a pool table especially developed for the client. The futuristic powder room is full of rhythm and mystery with its individually gilded and back-lit stone walls, the gold basin and luxurious Swarovski crystal taps.

Casa Forma 对这间 325 平方米的公寓的大胆和富于美感的设计，成功地打破了传统公寓中居住和娱乐不可兼得的局限。眨眼之间，精致的住宅变成一个奢华的娱乐空间。空间的趣味性通过室内独特的装饰材料、灯光、视觉效果和技术的应用，贯穿设计始终。

大厅里，首先映入参观者眼帘的是华美的地板和镶嵌在抛光黄铜框架上的玛瑙材质的透光顶棚镶板。24K 的金制台灯和定制的镶嵌着黄铜的意大利大理石地面，使整个空间都散发着一种 Casa Forma 设计所特有的舒适感。在该设计中，公寓的主体空间被扩大，并被分割成餐饮区、厨房和陈列着令人叹为观止的伦敦全景画的吧台区。餐饮区的背景墙是为该设计专门定制的，墙体是覆盖着金属花纹玻璃的引人注目的细木工艺品。定制的玻璃和黄铜材质的餐桌营造了令人愉悦的用餐氛围。

厨房区包含了一个宽敞的独立工作台和在其上方等长的精美的定制吊灯。当夜幕降临，独立工作台前方的区域便转换成一个有着发光玻璃地面的舞池。吧台区域则使用了多种优质的建筑材料，透光玛瑙板、深色原木、多孔无光泽的青铜制品和镜子。吧台区的 LED 屏幕和同步播放的音乐烘托出豪华私人俱乐部般的氛围。公寓左侧的走廊采用了手工黄铜叶镜面砖覆层的材质。叶子形的镜面瓷砖营造出一个迷宫，一路引领你到达台球厅。台球厅作为娱乐场所采用了水泡墙面并配置了为用户定制的台球桌。未来感十足的化妆间中，镀金的背光石墙、金色的水池和奢华的洛世奇水晶水龙头都使整个空间韵律感和神秘感十足。

LOCATION
Singapore District 11, Singapore

新英格兰风格住宅

NEW ENGLAND STYLE

DESIGNER *Heesoo Kang, Hege Torgersen*
DESIGN COMPANY *Design Intervention i.d.*
AREA 305 m²
PHOTOGRAPHER *Jo Ann Gamelo-Bernabe*

The interior brief, was to create a space that was light and classic in feel, but still looked fresh. They also had a collection of Peranakan items, artifacts and antique furniture that they wanted to incorporate into the space. The client also selected inspirational images that the designers incorporated in order to create their dream home. Their storage requirements were a major priority, so the designers designed built-in cabinetry planned down to the smallest detail taking into consideration the exact sizes of a DVD cover and CD player. The designers could not afford to waste any valuable space.

The furniture pieces such as the sofas, coffee table and side tables were also custom designed based on the flow through the house. We selected a palette of taupe and duck egg blue to achieve a calm and welcoming feel—yet giving a formality to the space in both the living and dining areas. The sophisticated

palette contrasts beautifully with the off-white cabinetry finishes and sets-off the cornice detailing that the designers used throughout the whole house. These shades were also used in the wallpaper that ran from the entrance wall, right through to the living and dining which helped connect these two areas and give complete synergy to the house.

The dining table was custom-made to fit the space of the narrow dining area perfectly. The challenge was to allow enough space to be able to walk through to the kitchen without it feeling like a corridor, so the chairs were also designed especially so as not take up more space than necessary but give the required distance. Even the beams in the dining area were constructed so that two lights and a fan would fit above the table without being out of proportion. The overall result is a complete house that looks calm, contemporary and fresh.

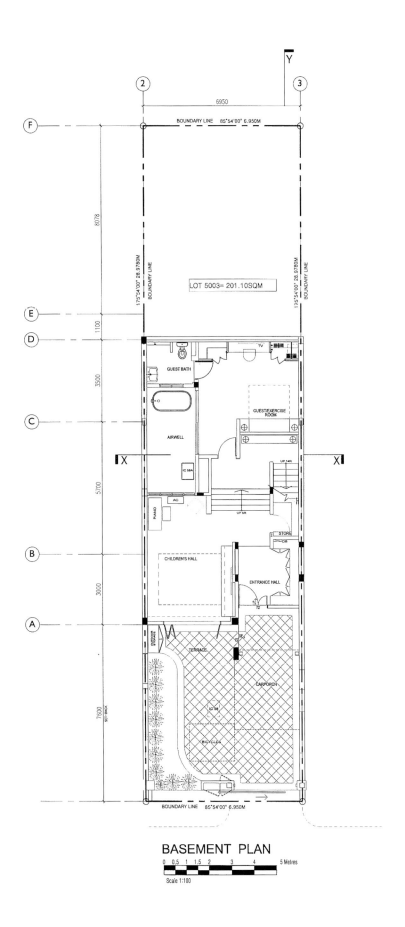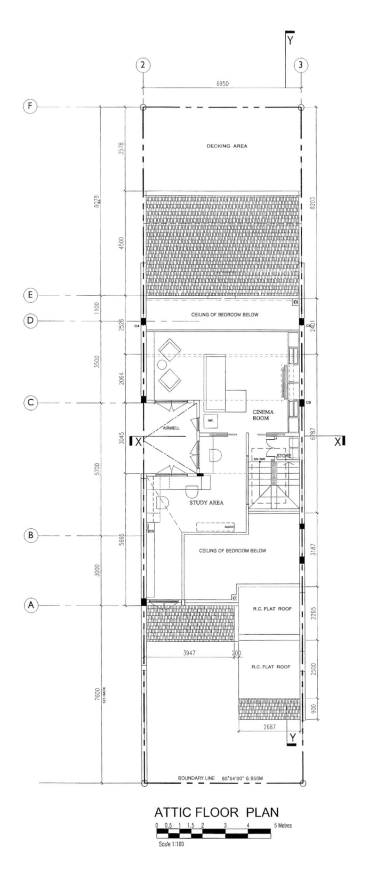

BASEMENT PLAN

ATTIC FLOOR PLAN

简单来说，该室内设计的目标是要打造一个浅色调且充满古典气息的空间，同时保留其清新的装饰风格。室内摆放了与整体设计风格相一致的住户收集的一些土生华人物品、手工艺品和古董家具。设计师加入了住户亲自选择的一些励志的图片，以便创造出他们理想中的家。住户的存储需求是首要考虑的问题，因此，为了不浪费任何可利用的空间，设计了内置橱柜。在设计过程中，设计师甚至考虑到了最微小的设计细节，如DVD外壳和CD机的具体尺寸。

室内的家具，如沙发、咖啡桌、小桌等也都是根据室内空间的流量大小定制的。设计师选择了灰褐色和鸭蛋蓝色的配色方案，以营造出一种宁静、热情的氛围，同时又使整个餐厅和起居室给人一种舒适、得体的感觉。精心设计的配色方案与乳白色喷漆的储物柜形成了完美的对比，并衬托了在室内广泛使用的檐板的细节设计。这些明暗度对比的搭配设计也被应用于墙纸上，墙纸从住宅入口处的门廊一直延伸至起居室和用餐区，不仅连接了这两个区域，也起到了协调室内整体装饰风格的作用。

定制餐桌摆放在狭小的用餐空间内正合适。而餐厅设计中最大的挑战就是如何腾出足够的空间，使人能够自由地活动于餐厅和厨房之间，同时又不会像走在走廊中那样拥挤。所以为了腾出所需要的空间，餐椅也是特别设计的，使它不占用多余的空间。空间中安装了横梁，以便使餐厅上方的两盏灯和风扇与餐桌相称。该设计最终呈现给大家一个氛围宁静、风格现代又清新的空间。

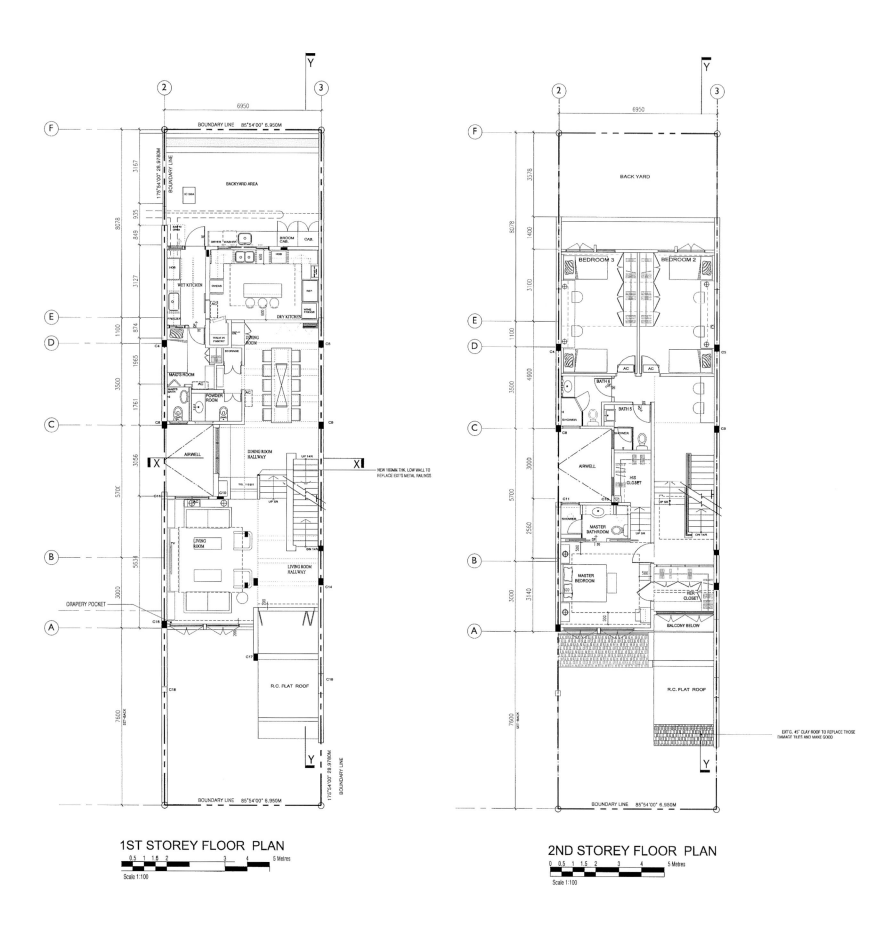

特色混搭

CHARACTERISTIC
& MIXMATCH

LOCATION
New York City, New York, USA

SOHO LOFT

DESIGNER *Laura Umansky*
DESIGN COMPANY *Laura U, Inc.*
AREA *279 m^2*
PHOTOGRAPHER *Barry Grossman*

A client purchased a New York penthouse apartment in the historic Singer Building on Broadway, and asked the designers to mildly renovate and completely furnish the home. The designers dubbed the style of this project ELECTRIC GRANDMA, based on the heavy mix of pattern and antiques, which was combined with a lot of vintage and antique furnishings.

The client had a very specific vision based on her favorite lodging in New York, the Crosby Hotel (just several blocks from the penthouse she ultimately purchased). The designers were asked to emulate the eclectic style of the Crosby throughout the client's home, as well as incorporate specially requested items the client had collected.

客户购买了这间位于纽约百老汇大街胜家大楼中的顶层公寓，并授权设计师对其进行适度的更新和全新的布置。根据室内复古风格的家具和古董家具的组合陈列所形成大量的图案和古玩的混合搭配，设计师把这一设计项目命名为"折中风格的老奶奶"。

客户最钟爱位于纽约的克罗斯比酒店的寓所的设计风格（克罗斯比酒店距离她最终购买的这间顶层公寓只相隔几个街区），她对房屋的室内装饰风格有着自己独特的品位和见解。按照她的要求，在整间公寓的设计中，设计师模仿了克罗斯比酒店的折中风格设计，并在室内空间的装饰摆放中，融入了客户要求放置的个人收藏的物品。

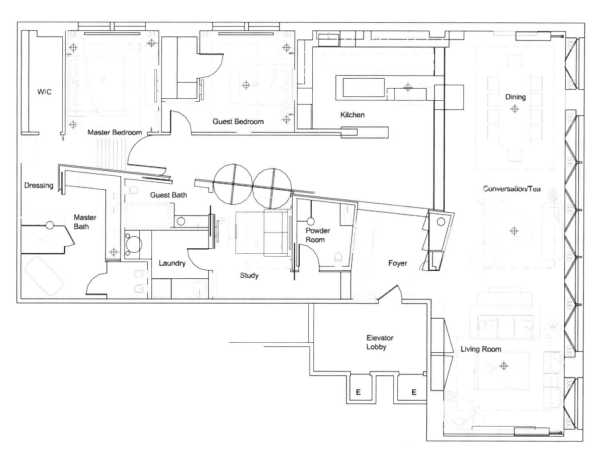

LOCATION
Hollywood Hills, California, USA

CLASSIC DESIGN

DESIGNER *Jennifer Dyer*
DESIGN COMPANY   Jeneration Interiors
AREA   232 m²
PHOTOGRAPHER   G Crawford

洛杉矶经典公寓

This beautiful home is perched high in the Hollywood Hills. The home is 232 m² on the three-split levels. Terraces span the length of the home and allow for breathtaking views of the Los Angeles skyline and the Pacific Ocean. The home has two bedrooms and two bathrooms with a private separate room down the hillside. This room could be used for a gym or office. The master bedroom walls are beautiful, deep shale gray with pops of kelly green accents.

The airy living room and dining room are mostly white with various textures, including linen and cowhide, which create an eclectic look. The family room has upholstered wall panels in black and white. Lastly, the dining room opens one two sides to a luxurious garden with a fountain allowing guests, when dining, to enjoy the mild Southern California weather.

这座精美的房屋坐落在好莱坞山上。室内面积为 232 平方米，共分为三个部分。在环绕整个房屋的露台上可以尽情欣赏壮观的洛杉矶天际线和太平洋美景。室内有两个卧室和两个盥洗室。山腰上还有一间独立的私人房间，可用于办公或者健身。主卧室的墙壁是华美且带有黄绿色斑点的深页岩灰。通风良好的起居室和餐厅以白色调为主，但通过使用亚麻、牛皮等不同的材质，营造出不拘一格的风格。家庭房中使用了黑白相间的软体墙板。餐厅两侧敞开，面向奢华且带有喷泉的花园，使客人在用餐的同时，尽情享受南加州温和的天气。

LOCATION
Venice, California, USA

威尼斯小屋

VENICE BUNGALOW

DESIGNER *Jamie Bush*

DESIGN COMPANY Jamie Bush & Co.
AREA 111 m²
PHOTOGRAPHER Laura Hull

By blurring interior and exterior areas, making premeditated plays on scale, and adding just-so doses of color and texture, Bush pulls off a real-deal decorating feat: He doesn't just make small feel bigger. He manages to make a diminutive 1950's box feel downright opulent. Defining and unifying the rooms is key to the project. Bush's architectural background reveals itself in his approach to elongating the house by aligning interior doors, so that both the width and depth of the space are broadened by unfettered sightlines. Stand in his library, with its 3-m Futurama in Los Angeles sofa and gallery-size abstract painting, and you can see all the way through the music room and the living room to the hallway.

Bush also draws the eye up by wallpapering many of the ceilings. In the bedroom, he uses a grain-like Keith McCoy paper. In the library and music room, it's a custom McCoy fern pattern. He also "wraps" rooms in tile, fabric and wood. And for the library, Bush's favorite room in the house, he took the bleached white oak he used for the floors and extended it up onto the walls and built-ins for a "Japanese sauna box effect". Similarly, for the bedroom, Bush took a few bolts of "faded lavender" Belgian linen and draped it along almost every vertical surface, effectively covering every square inch (the closet, the bookshelves, the windows) in yard after yard of ripple-fold fabric. The lone exposed wall is painted in a custom-blended Pratt & Lambert paint color that seamlessly matches the drapery.

Bush's other design mantra for small spaces: repetition: the eight identical carriage house French doors in the living room, library and music room; the dining room's matching mirrors; the twin Karastan carpets in the library and music room; the sofas in the living room sofas. Finally, by enclosing the entire house in 2.5 m high walls and ficus hedges, Bush creates a sense of privacy, while also increasing the square footage of the living space.

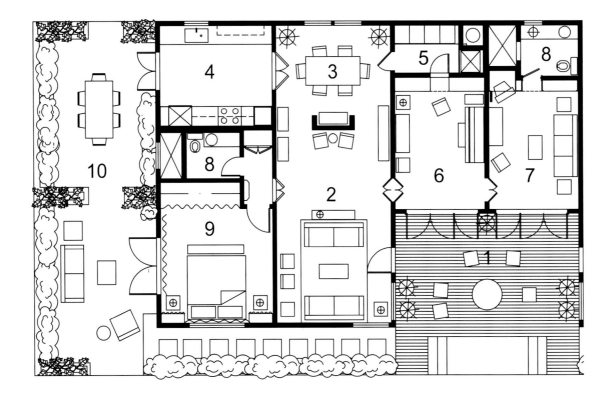

1. Front Deck
2. Living
3. Dining
4. Kitchen
5. Pantry
6. Studio
7. Den
8. Bathroom
9. Master Bedroom
10. Back Patio

通过弱化和模糊室内和室外区域的界限，设计师 Bush 依照预先的设计计划，紧锣密鼓地在空间内开展装饰设计，细心地加入了少量的色彩和纹理，努力完成了这一真正意义上的装饰设计壮举。他所做的不仅仅是使小的东西看起来更大，而是使一个很小的 20 世纪 50 年代的盒状建筑变得丰盈。界定并整合住宅中的房间是该设计的重点。将室内空间中的门排列成行，使空间得以扩展，并使整个空间的宽度和深度在毫无阻碍的视线中得以延伸，这也呈现出了建筑背景。站在图书室内，身边是 3 米长的 Futurama in Los Angeles 沙发和画展尺寸大小的抽象油画。视线可以穿过室内的音乐室和起居室，一直延伸到房屋的走廊。

通过给许多空间中的天花板贴上壁纸，设计师把观赏者的注意力吸引到室内的上方空间，同时为卧室选择了颗粒状的 Keith McCoy 墙纸。在图书室和音乐室内，选择了 McCoy 蕨纹壁纸。除此之外，设计师还为房屋穿上了瓷砖、布料和木质"外衣"。图书室是设计师最喜爱的房间，他沿用了其他空间中地板的材质和漂白的橡木，将其进一

步应用在墙壁及其内置部分中，打造出如同日式桑拿浴房般的效果。同样地，设计师为卧室选择了几匹如同褪色了的薰衣草淡紫色比利时亚麻布，它们被应用于几乎所有的垂直表面上，这种波纹表面的布帘覆盖了每一平方米的表面（壁橱表面、书架表面和窗户表面）。唯一的外墙则粉刷了 Pratt & Lambert 定制有色涂料，与淡紫色布帘形成完美的搭配。

设计师的另一个针对小空间设计的"咒语"是"重复"。无论是起居室、图书室和音乐室中那八个相同的法式落地双扇玻璃马车门，餐厅中的镜子，图书室和音乐室中分别摆放的一对 Karastan 品牌地毯，还是起居室中的沙发，都体现了重复性。最后，设计师以 2.5 米高的墙壁和榕树篱笆将整个房屋围住，从而营造出一种私密空间的氛围，同时也扩大了居住空间的建筑面积。

LOCATION
Connecticut, USA
WILTON RESIDENCE
DESIGNER UXUS
AREA 494 m²
PHOTOGRAPHER *Dim Balsem*

Wilton 住宅改建

The Wilton Residence is a 1930's hunting lodge located in the historical heart of Connecticut. The home belongs to an American family, who after living in Europe for 13 years returned to the US. The interior, a reflection of their years living abroad, is an eclectic mix of high European contemporary design, personal mementos and a classic American style.

UXUS wants to create an invigorating energy by fusing a pioneering spirit with the inherent comfort of effortless style—colonial yet innovative. Inspired by Rustic Luxury and essential beauty found in simple things, every space within the Wilton Residence exudes openness, personality and playful sophistication. When designing the spaces, the key objective is to create an inviting environment for the family and their guests. Interior spaces celebrate the family's stories, moments and inspirations with a whimsical touch. The Hunting lodge is remodeled to open the space and create a contemporary living environment where spaces and activities flow together yet maintain the heritage of the home's original use—a hunting lodge.

The furnishings and décor vary from family momentos, antiques, and vintage furniture to high contemporary Dutch and American design. Unique pieces, such as re-purposed chairs decorated with vintage oil paintings, create a show-stopping exhibition. A "Not So Fragile" table with a fluorescent orange accent animates the living room. The family room features towering bookcases that form magical columns, creating the illusion of hovering books.

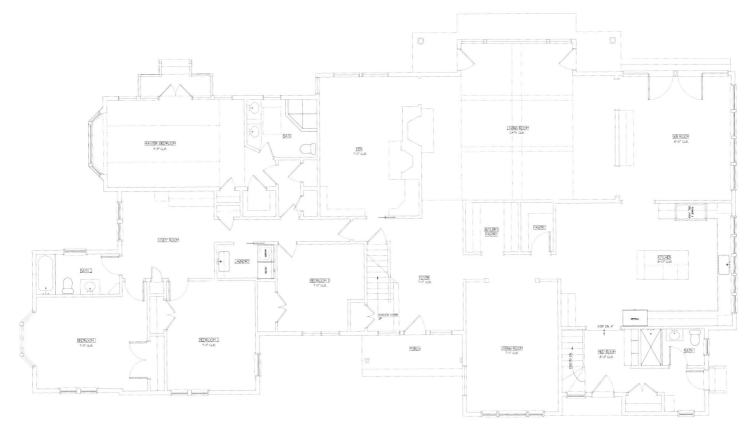

FIRST FLOOR PLAN
SCALE: 1/4" = 1'-0"

219

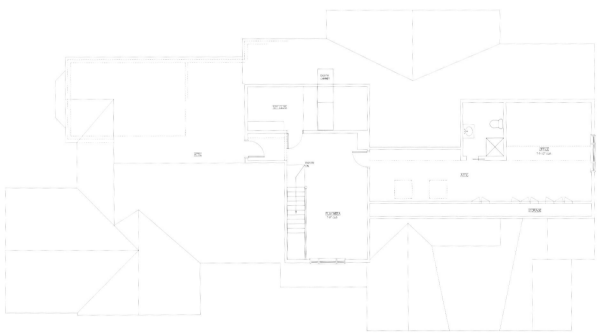

　　Wilton住宅原是一个建于20世纪30年代、曾位于康涅狄格州中心区域的一个狩猎小屋。这座小屋属于一个在欧洲生活了13年之后重返故土的美国家庭。室内的装饰设计是他们多年海外生活的写照，折中混合了高水准的现代欧洲设计风格和经典的美式设计风格。

　　UXUS设计公司想要通过结合室内设计的开拓精神和住宅中固有的舒适的随性风格（原有的殖民地风格和创新的设计风格），在整体空间内彰显出令人振奋的活力。从简单的事物中所具有的野趣奢华和内在美感中得到启发，每个宽敞的空间都散发着独特的个性并体现了趣味装饰的复杂性。设计师的首要目标便是要为居住者和来访者创造一个极具魅力的环境。空间中以奇异的装饰记录了整个家庭的故事，所经历的重要时刻和鼓舞人心的事件。在保留了空间的原始功能——作为一个狩猎小屋的同时，设计师对房屋进行了改造，力求使空间现代感十足。

　　起居室的风格由家庭的回忆、古董和复古的家具变换到高端的现代荷兰和美式设计风格。独一无二的装饰，例如复古风格的油画装饰着由回收材质制成的椅子，如同引人注目的展览品。一张"不那么易碎"的荧光橙色桌子，使整个起居室充满活力。家庭室的特色是其摆放的高大的书架，如同神奇的圆柱，使人产生书籍在空中盘旋的错觉。

LOCATION
São Paulo, Brazil
TERRACE 2

DESIGNER *Fabio Galeazzo*
DESIGN COMPANY *Galeazzo Design*
AREA 88 m²
PHOTOGRAPHER *Lufe Gomes*

二号平台屋顶式建筑

With 88 m², the space has a lot of colors and its inspiration comes from the crossing of the east (orient) and west (occident) cultures, in a contemporary way.
With the segmented painting on the walls and roof, the designer aims to break the place amplitude in a bold and irreverent way, which reminds the origami's folding.
Bamboo furniture, lacquer and hardwood, traditional references to China flow over the high-tech bamboo floor and contrast with the extravagant wallpaper printed with stylized golden dragons.
Important occident pieces fulfill the ambient, such as the crystal chandelier from the traditional French brand Saint Luis, the 1950's wall lamp from Serge Mouille, a circular shaped shelf and a huge rough linen sofá placed in front of a big fireplace made of exotic granite.

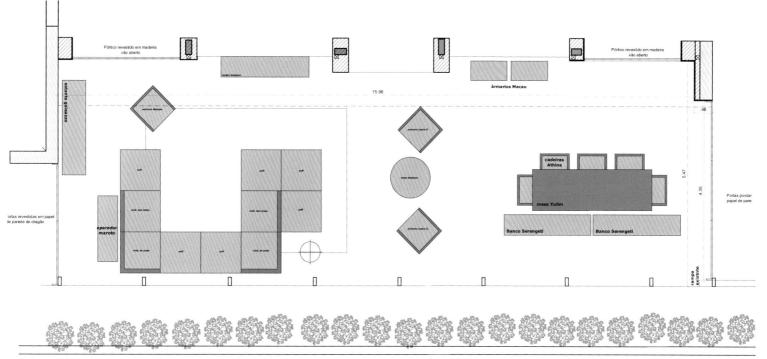

这座面积为88平方米的住宅设计融入了很多色彩搭配,这一设计灵感来自于东西方文化在一种现代的方式下的交融。墙壁和屋顶上摆放着组合套画油画,体现了设计师想要以一种大胆且非传统的方式,如同手工折纸一般来分割这个宽敞的室内空间。

高科技的仿真竹制地板上摆放着竹制家具、漆器和硬木制品,这些都是中式装饰风格的写照,并与奢华风格的金色龙图案的壁纸形成对比。

在整体装饰中,重要的西欧风格装饰品贯穿在整个空间中,例如来自历史悠久的法国传统品牌 Saint Luis 的枝形吊灯,20世纪50年代风格的 Serge Mouille 壁灯,环形书架和摆放在充满异域风情的花岗岩材质壁炉前方的巨大的粗亚麻材质沙发。

LOCATION
California, USA

HOLLYWOOD HILLS RESIDENCE

DESIGNER *Elizabeth Gordon*
DESIGN COMPANY *Elizabeth Gordon Studio*
AREA *474 m²*
PHOTOGRAPHER *Grey Crawford*

好莱坞山住宅

Elizabeth designed the interiors of the colonial brick and stucco home using both antique and modern references in the furnishings. She utilized tiny antique shops and flea markets to upper-crust stores and showrooms in LA, Palm Springs, and New York, mixing the quirky with the sophisticated in defining the aesthetic. When Elizabeth couldn't find the exact item she wanted for a room, she designed it herself and had it custom made. In the family room, Elizabeth created a custom media cabinet with a secret door that opens to a media room beyond. The room, equipped with plush sofa, cork fabric-wrapped ottomans, graphic print chairs, vintage lamps and hide rugs, created a cozy, intimate space that was perfect for the family to relax together and to entertain. As with all of the spaces, there was an element of surprise—a graphic print wallpaper at the ceiling—that keeps the space feeling fresh and modern. In the dining room, Elizabeth took a cue from the oversize windows looking out onto the greenery of the backyard and decided to play on the views she found there. Dressing the room in a color palette of

cool greens and ebonized wood, Elizabeth topped the cut sea-grass rug with a custom French-polished wood table, a custom ebonized wood sideboard decorated with vintage bronze pulls, vintage lamps, and high-backed chairs. The modernizing touches to the space—a vibrant-colored commissioned painting, a custom designed faux-bois metal chandelier, and wallpaper at both the walls and ceiling—allow the room a comfortable formality perfect for evening gatherings over dinner. Throughout the home, Elizabeth pairs together a mix of furnishings and architectural details that play on color and materials, balancing glamour and formality with warmth and whimsy. The result is a harmonious eclecticism that embraces the idiosyncrasies of the client's family life and reflects her own unique tastes.

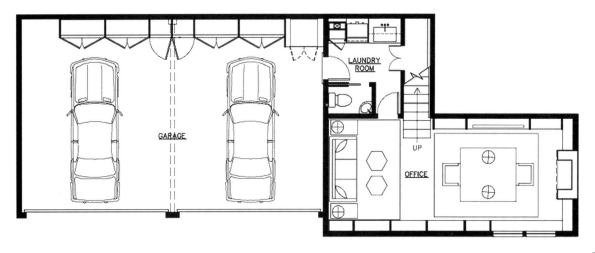

Basement Floor Plan

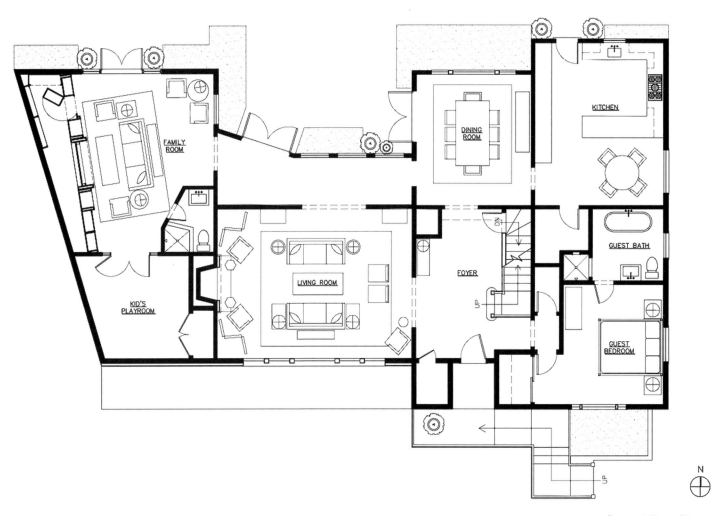

Ground Floor Plan

在对这间粉刷过的殖民地风格砖瓦房屋进行室内装饰时，设计师 Elizabeth 同时使用了老式和现代的装饰元素。在装饰过程中，她充分利用小的古董商店、跳蚤市场，同时也造访了高档的商店和位于洛杉矶、加州棕榈泉和纽约的装饰品展厅。室内空间混合搭配了奇异、高雅、精致的设计元素。

如果在装饰品商店中不能找到适合室内风格的装饰品，Elizabeth 便自己设计并将其定制出来。在家庭室中，Elizabeth 设计了一款定制的多媒体组合柜，并附带有一个隐形门，通向后方的媒体室。空间中摆放了绒毛面料的沙发，用面料包裹的软木材质搁脚凳，印有几何图案的椅子，老式的灯饰和隐形地毯，共同创造出一个供家庭聚会的娱乐、放松的理想场所，并使整个空间充满了活力和温馨之感。所有空间中都有一个令人惊喜的装饰元素，那就是天花板上铺设了印有几何图形的墙纸，提升了整个空间的清新感和现代感。在餐厅中，Elizabeth 从超大型的面向后院绿色植被的窗户中得到了提示，并决定充分利用这个景观。Elizabeth 为室内空间选择了冷艳绿色的配色方案，并使用了乌木色木料。Elizabeth 选择了整齐的海草地毯搭配定制的抛光木桌和搭配有老式的青铜拉手、老式的灯饰和高背椅的乌木色木质定制餐具橱。

现代风格的室内装饰包括：一件色彩鲜亮夺目的定制油画作品，特别设计的仿木纹金属枝形吊灯和铺设了壁纸的墙壁和天花板，使整个空间成为餐后家人聚会的理想场所，让人倍感舒适。在装饰布置中，Elizabeth 使室内家具的布置搭配建筑细节，注重配色方案和装饰材料的选用。通过温馨和奇异的设计，平衡了整个空间的庄重感，最终呈现出一个和谐的折中主义设计典范，既体现了住户一家特有的风格，同时也很好地展现了设计师自身独特的品位和设计审美。

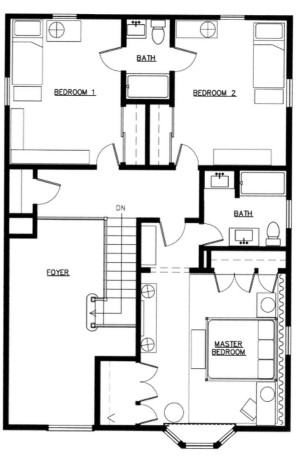

2nd Floor Plan

LOCATION
Moscow, Russia

PRIVATE HOUSE IN MOSCOW AREA

DESIGNER *Roman Leonidov, Olga Budennaya*
DESIGN COMPANY   Leonidov and Partners
AREA   730 m^2
PHOTOGRAPHER   Alexey Knyazev

莫斯科地区私宅

By the layout, this house reminds of a catamaran: two hulls of the private households are joined by a double-lighted living-room "deck". However both hulls are turned upside down and form impressive consoles which gives the house a very dynamic outline.

The asymmetry of the main façade is artfully emphasized by the materials employed: for the most part, including the entrance portal, it is decorated by the light stone, while the advanced volume is faced by the brown wood. Wide verges are hemmed with wood as well. The picture is capped by the terraces—one of them rests on the wide pillars shaping entrance to the house, while the other is supported by the thin metal props. The ground (first) floor of the house is dedicated to the public areas, while the first (second) floor hosts the master bedroom and children's premises including one playroom. Offsetting of the main entrance in respect to the central axis allowed for the spacious double-lighted living-room to become a center of the whole composition. It is separated from the entrance zone by the staircase connecting all floors of the house including the basement level which hosts the home movie-theater. For the purpose of separating the stairwell hall without turning it into the blind box, the architects have come up with an idea of using the glass partitions with the elegant floral ornament matted on. Up-close the pattern impresses with its richness and exuberance and from a distance it reminds you of the frost drawings and makes you imagine the streaming water on the surface of the glass.

通过设计，整栋私人住宅的外观使人联想起双体船：两个独立的私人住宅由一个双倍照明的起居室（好像一个甲板）连接起来。这两栋建筑颠倒的外形共同构成了一个引人注目的支柱，使整个连体住宅看起来活力十足。

建筑材料的使用，突出了整栋建筑正面的不对称外观。住宅的大部分空间，包括入口处的大门，都使用浅色石料进行装饰。而其正前方的空间内，则使用褐色装饰木料。宽阔的屋顶端边包裹着木边儿。平台屋顶下挂着装饰画，其中一个平台屋顶放置在住宅入口处的宽大的柱子上，另一个由一根细的金属柱子支撑。底层主要作为公用区域，二楼则包括主卧室，儿童房间和一个游戏房。大门不在整栋建筑的中央，偏离了中心轴，这样设计是为了凸显双倍照明的起居室在整个房屋组合中的中心位置。起居室通过连接住宅中每一层的楼梯（也连接了设置有家庭影院的地下室）与入口区域相分离。为了独立划分出楼梯间，不把它变成一个封闭的如同百叶窗匣的狭小空间，设计师使用了玻璃隔墙并在上面铺设了精美的花朵图案的装饰。近距离看，装饰物上的花朵繁茂、丰富；远距离看，整个图案会让人联想起描绘霜花的图画，或者透明玻璃上的流动的水珠。

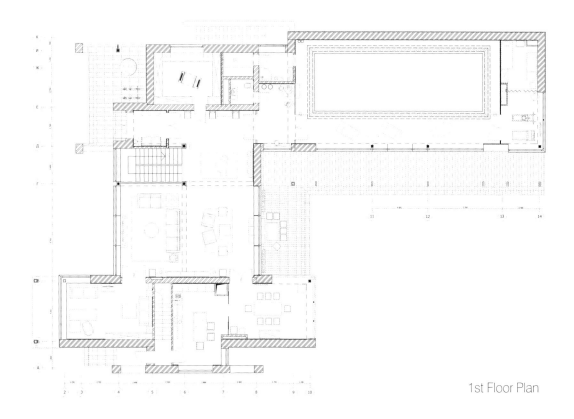

1st Floor Plan

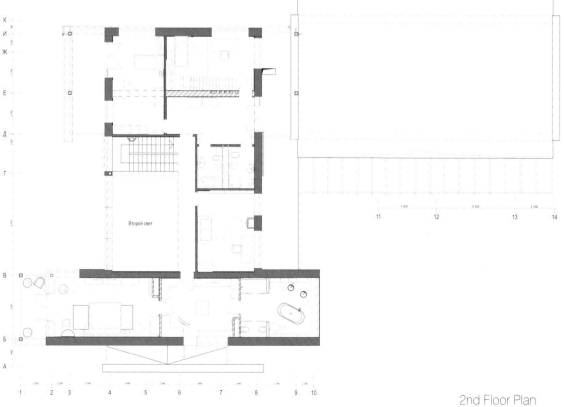

2nd Floor Plan

现代简约
MODERN
& SIMPLE

LOCATION
São Paulo, Brazil

城市森林

URBAN FOREST

DESIGNER Fabio Galeazzo
DESIGN COMPANY   Galeazzo Design
AREA   300 m^2
PHOTOGRAPHER   Marco Antônio

The large amount of natural light allowed the abusive use of coating materials such as wooden walls of sustainable management in balsam color and cement floor burned, conferring modernity and boldness to the place. The property has been decorated with a mix of furniture design contemporary in contrast to raw wood on the walls and furniture distributed along to the environment.

In the living room, there is an angled couch which was designed by the office and received several pads of different textures. Tord Boontje armchair marks its presence in the room where dual banking logs floats near to the huge red carpet. In the black glass dining room table, there is a sculpture by Gaetano Pesce, and at the bottom of the frame a hyperrealist paint by Rodolfo Valdez surprising perspective. The final touch is up to the trio of black and gold chandeliers and the conceptual centerpiece of the Italian designer Gaetano Pesce.

The master bedroom interior reveals the whole vision in tune with the house. Details in the decoration prevail welcoming details as the head covered with straw silk. The striped bedspread and the African frame on the wall united with the Ingo Mauer lamp bring the boldness of the living but with a cleaner touch. Spaciously, the room is integrated through the bathroom door mirror large. The bathroom has concrete used in walls, floors and countertops which give us a contrast and draws attention to the synthetic marble.

　　室内充足的自然光照，使设计师可以尽情地使用可持续性环保香脂色木质墙壁和烧制水泥地面等，赋予整个空间一种现代感和醒目感。整栋住宅使用了现代风格家具，并与原木材质墙壁和分散摆放的原木材质家具形成对比。

　　起居室中摆放着由设计公司设计的以不同的纺织材质作为垫衬的斜角沙发。空间中摆放着 Tord Boontje 扶手椅，巨大的红色地毯旁"漂浮"着两片筑堤原木。餐厅内，黑色玻璃材质的餐桌上摆放着由 Gaetano Pesce 创作的雕塑，框架底部是由 Rodolfo Valdez 以惊人的角度创作的高度写实主义油画。最后一组装饰是三盏由黑色和金色组合的垂吊灯，以及由意大利设计师创作的概念艺术作品。

　　主卧室的装饰风格与整体空间的装饰风格相一致。细节设计得到了极致的发挥，如床头部分包裹着稻草色丝绸。条纹的床罩和挂在墙壁上的非洲风格框架与 Ingo Mauer 灯饰相搭配，以一种清新的装饰手法，彰显出狂野的生活方式。房间的宽敞通过浴室门上的大镜子凸显出来。浴室内的墙壁、地面和台面板使用混凝土的表面，在视觉上形成对比，也凸显出空间内的合成大理石。

LOCATION
Günesli, Istanbul, Turkey

GUNESLI PARK GARDENYA

DESIGNER *Meltem Çakmak, Engin Kurşit*
DESIGN COMPANY *Pebbledesign / Çakıltasları Mimarlık Tasarım*
AREA *150 m² (3+1 Sample House)*
PHOTOGRAPHER *Murat Tekin*

In the living room, the main concept was Scandinavian style and the different four season forms of a leaf. The angular lines used at the floor covering and the furniture, helped to show the last users a different point of view and how differences could be made. Because of the glass façade on one side, the TV unit was situated in the middle and the sofa was situated in front of the long wall which gave every seat the chance to observe the view. With the natural materials and patchwork carpet on the floor, the designers aimed to have a warm living space while trying to add some movement with the wooden frames used on the walls. The wall clock on the left wall at the entrance was designed by Pebbledesign, following our general "leaf" concept.

At the kitchen cupboards, wood and natural colors made a calm space where the whole family gathered. With this natural space, the designers used red in the accessories to avoid monotony. For the bedrooms, different concepts were formed according to the family scenario that was created at the beginning. These different concepts with different colors and materials enriched the project, which was based on natural colors at the living areas. In the master bedroom, the designers used natural wood sticks and mirrors to create linear designs in relation with each other. Several uses of mirrors gave different perspectives and several uses of hats as accessories and also lights, helped to give hints about our imaginary characters' lives.

　　公寓的起居室以斯堪的纳维亚设计风格为主，体现了叶子的形态在四季中的变化。地面和家具使用的配饰从另一个角度展示了不同的空间装饰风格是如何形成的。由于房间的一侧使用了玻璃墙，电视柜被摆放在了室内的中央位置，而沙发则被放置在长墙的前方，以便坐在沙发上不同位置的人都能够观看电视上的画面。室内使用了天然材质的地板和拼接地毯，以营造出一个温馨的起居空间。同时也给墙上的木框增加了一些设计细节，以突出其特点。入口处左侧墙壁上的挂钟是由 Pebble Design 设计的，其风格与"叶子"的主题设计理念相符。

　　厨房中，橱柜采用木色和自然色相结合的色彩方案，营造出一个全家人相聚在一起的宁静而温馨的空间。在这个充满自然气息的空间内，厨房中配饰品的颜色则选择了红色，以避免空间色彩的单调。在卧室的设计中，根据最初创造的家庭场景融入了不同的设计理念。这些不同的设计理念、配色方案和各异的装饰材料丰富了以生活区中自然的配色方案为基础的整体设计。主卧室中，天然的木条和镜子均为线形设计，相互呼应。室内几处镜子的使用，为空间提供了不同的视觉角度，还有室内摆放的几顶装饰性帽子和照明设计，都隐约地展示了设计师脑海中虚拟人物的生活。

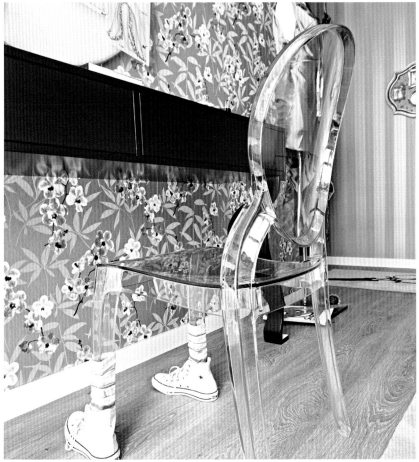

LOCATION
Loughton, London, UK

FIVE BEDROOM DETACHED FAMILY HOME

五间卧室的独立住宅

DESIGNER *Ravi Lakhaney, Gurjeet Hunjan*
DESIGN COMPANY Boscolo
PHOTOGRAPHER Christina Bull

First, the designers extended the property at the rear to facilitate a spacious second sitting room. The designers also knocked through a wall separating the existing kitchen and utility room creating a substantial kitchen-diner. The designers made extensive use of stone tiles, using a slab finish. A tranquil, earthy color palette was used throughout, with the emphasis on naturally occurring shades, but with a luxurious finish and detailing subtly enhancing visitors' experience.

The centerpiece of the design ethic, in a home that was about subtlety not centerpiece, was reached in the dining room which is dominated by a piece of artwork we sourced and manipulated to suit the space. The designers played off the picture using a stone dining table, keeping with the natural but luxurious aesthetic, the ripples in the stone subtly echoing the rocky paths in the picture.

Subtlety of experience meant carefully considering the harmony of all the elements within the home, none more so than the lighting. With a natural, flat palette of colors used for the walls and soft-furnishings, there was a danger of understatement becoming lack of statement and we felt lighting was the key to preventing this. To this effect, the wall art in the dining room was washed by down lights ensuring prominence, when the rest of the room's lighting was delivering an intimate feel. The bedrooms used table lighting rather than down lights to create a tranquil effect conducive to relaxation. The designers used layers of light in the kitchen with down lights, wall lights—washing light up and down—and pendant lights over the breakfast area, these being primarily decorative, with a concrete finish to continue the natural look. With the kitchen space being light and bright and dominated by clean lines, the designers also added a bit of chaos with Philippe Starck Masters chairs in black to break up the order and add depth.

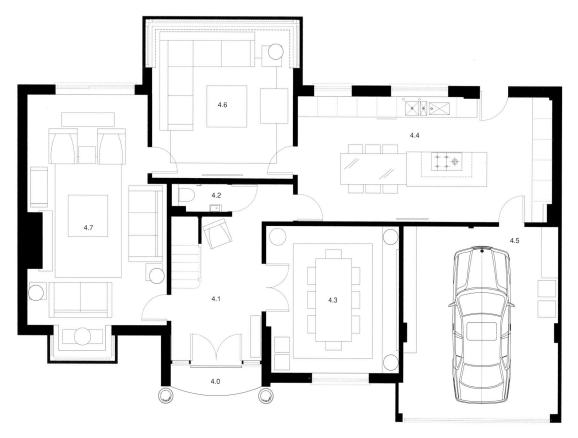

PROPOSED GROUND FLOOR PLAN

4.0   FRONT EXTERIOR
4.1   HALLWAY
4.2   GUEST WC
4.3   DINING ROOM
4.4   KITCHEN
4.5   GARAGE
4.6   LOUNGE
4.7   LIVING ROOM

在该设计改造中，设计师首先扩展了住宅后方的空间，并将这个空间改造成房屋中的第二个客厅。同时拆除了厨房和杂物间之间的墙壁，并将整个区域改造成一个开放式厨房。在装饰过程中，板式石质贴砖也得到了广泛的应用。整个房屋选择了宁静、质朴的配色方案，以突出在自然状态下室内空间的明暗变化，但同时进行了细节的点缀并配以华美的涂饰，以增强观赏者的体验感。

该设计想要传达的设计理念是：家居设计的精髓在于室内的装饰细节，而不是室内的中心装饰品。这一设计理念在配有一件与其风格相符的艺术品的餐厅的设计中得到了很好的体现。为了与空间中的艺术画交相辉映，设计师为餐厅选择了一张石质的餐桌。石桌天然的波浪似的石纹隐约地与艺术画中崎岖的石路相互呼应，使整个空间的自然、华美的风格得以延续。

设计中强调的细腻的体验就是要考虑室内空间中所有设计元素之间是否相和谐。从这个角度讲，灯光设计是最为重要的。由于墙壁选用了自然的浅色调，而且室内空间使用了软装饰材料，这种低调、内敛的风格有可能会使观赏者觉得整个空间缺乏鲜明的特点，而灯光设计在防止观赏者产生这样的错觉中，起到了至关重要的作用。为了突出整体空间的设计风格，餐厅内的艺术墙壁使用了下照灯，使其得以凸显，而其他灯光则给人一种温馨的感觉。卧室没有使用下照灯而是桌灯，以营造出一种宁静、放松的氛围。厨房中使用了叠层灯光，墙壁上、下端使用了下照灯。早餐区的上方安装了吊灯，作为最基本的装饰性家具，吊灯的表面使用了混凝土涂料，以延续室内空间的自然风格。厨房区主要使用了流线设计和同样风格的照明，为了给空间增加一些"混乱感"，设计师在其中放置了一把由 Philippe Starck 设计的黑色 Masters 椅，以打破整个空间原有的秩序，并提升整体设计的内涵和深度。

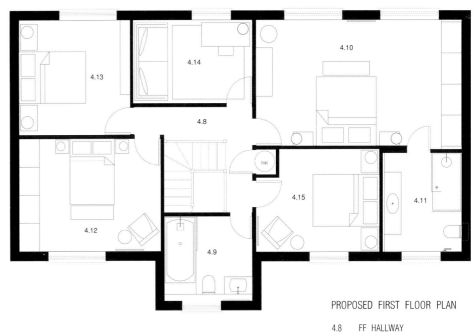

PROPOSED FIRST FLOOR PLAN

- 4.8 FF HALLWAY
- 4.9 FAMILY BATHROOM
- 4.10 MASTER BEDROOM
- 4.11 MASTER EN-SUITE
- 4.12 BEDROOM 1
- 4.13 BEDROOM 2
- 4.14 BEDROOM 3
- 4.15 BEDROOM 4

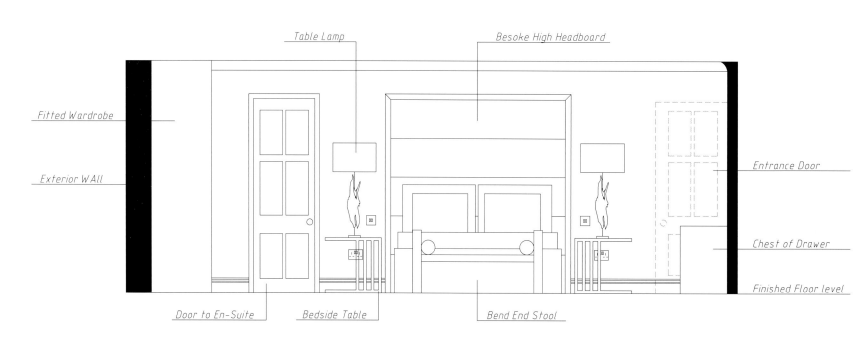

Master Bedroom Elevation

LOCATION
Fisher Island, Florida, USA

CRITZ RESIDENCE

Critz 住宅

DESIGNER *Pepe Calderin*
DESIGN COMPANY   Pepe Calderin Design
AREA   418 m²
PHOTOGRAPHER   Pep Escoda

Modern, sleek and chic are additional adjectives that apply. Calderin opted for European lines to make his eye candy statement through furniture and appointments, most of which have artistic overtones. Individual elements stand on their own and work together at the same time. One of the residence's most irresistible rooms is the breakfast nook where Mademoiselle chair upholstered in a floral fabric by Missoni surround a black lacquer table.

Nearby, corkscrew-like aluminum barstools line up to the bar. A metallic pendant lamp that resembles an origami sculpture further perpetuates the room's whimsy, a tone solidifed by the undulating white brick bookcase that holds the homeowners' many mementos and toy-like art.

More serious in tone is the dining room where a custom table in black lacquer nods to Art Deco's faceted forms and trapezoidal qualities. Jeanette chairs by

the Campana Brothers are positioned to be enjoyed visually, not so much to be sat on. An Autrian crystal chandelier with clear, yellow and green pendants hangs like an art installation across the ceiling.

In the living room, the furniture equation offers a dichotomy of modernist style by leading European designers: a milk-white modular sofa by Francesco Binfare, an iconic Up 5 chair by Gaetano Pesce, a Screw side table by Euro Aarnio and the Rainbow chair by Patrick Norguet. The whimsy even finds its way into the powder room with Bisazza mosaic in swirling patterns of gold andblue and a hatbox lavatory.

"Even though I've done 'white' rooms before, nothing comes close to this", says Calderin. "It was a great, fresh departure for me."

我们可以用现代、时髦、新潮这些形容词来描述这间公寓。Calderin 选择了欧式风格的线条设计，通过室内富于美感的家具及陈设来诠释其华美、悦目的设计风格。公寓中各个设计元素均有其自身的特点和价值，同时又相互配合，共同构成室内空间的整体装饰风格。公寓中有许多设计精美的空间，早餐区便是其中之一。黑色喷漆的餐桌周围摆放了以 Missoni 设计的 Mademoiselle 面料为花卉图案的装套座椅。旁边的吧台前则排放着螺旋形的铝制酒吧高脚凳。仿纸质手工雕塑的金属材质吊灯延续了古灵精怪的设计风格。摆放着主人的纪念品和玩具艺术品的白色波状表面的砖制书架是对该风格的进一步强化。

餐厅的装饰风格更加严肃，定制的黑色喷漆餐桌是向 Art-Deco 建筑风格中的不规则多面体外形设计创意的致敬。空间中摆放着由 Campana Brothers 设计的装饰性 Jeanette 餐椅。Autrian 水晶枝形吊灯晶莹剔透的黄色和绿色的灯坠悬在半空中，如同安装在天花板上的艺术装置。

在起居室内，由欧洲顶尖设计师设计的家具以均衡的布局摆放，是对现代装饰风格的另类诠释，例如，由 Francesco Binfare 设计的乳白色组合沙发，由 Gaetano Pesce 设计的经典 Up 5 座椅，由 Eero Aarnio 设计的螺丝钉造型的边桌，由 Patrick Norguet 设计的彩虹条纹座椅。盥洗室中铺设的蓝色和金色相嵌的漩涡图案的碧莎马赛克和帽盒形马桶也彰显出设计风格的古灵精怪。

"尽管以前我也做过'白色'系列的室内装饰，但是以往设计的效果远比不上这一次，"设计师 Calderin 自己说道，"本次设计对我来说具有非比寻常的意义，在我的设计生涯中是一个具有里程碑意义的全新开始。"

The apartment of 145 m² was almost in the last phase of finishing process when the young Brazilian architect met the owners. The couple was displeased with the environment arrangements and the traditional decoration, so they gave the architect free hand to reformulate everything.

The base for the decoration was the fusion of polymer cement, minty and blue colors and the use of a whole wall covered in pinus, a wood that has virtually no commercial value in Brazil.

Another high point of the project was to expose the structural column right in the entrance door, composing a beautiful set with the dinner table. An important point was: all the main pieces such as tables, sideboards and sofa are designed by the architect. The owner was already a great admirer of Guilherme's work, so during the whole process she insisted on buying several pieces designed by the architect.

This partnership worked out so well that she decided to open a company just to represent Guilherme Torres's products in all Brazil and around the world. Other design pieces that made a difference were the Shadowy chaise, Moroso, Guilherme decided to use inside the apartment, since it was originally an outdoor piece. The rugs and wooden chair with steel legs were from Diesel and the dinner table chairs were from Driade. The piece of art beside Shadowy chaise was by João Machado and the one above the sideboard is by Eliane Prolik. In the master bedroom, the decoration was completely the opposite to the explosion of colors of the living room. Black and white used on furniture, bed linens and even the frames. Above the headboard, there was an art piece of João Machado and bed linen of Ari Beraldin. The kitchen received furniture in black lacquer, countertops in Corian and Smeg home appliances. The living room walls, gave continuity to the kitchen, since there are no doors dividing the spaces.

RL 住宅

RL HOUSE

LOCATION *Curitiba, Brazil*

DESIGNER *Studio Guilherme Torres*

AREA 145 m²

PHOTOGRAPHER *Denilson Machado of MCA Estúdio*

RL HOUSE
FLOOR PLAN

当这座面积为 145 平方米的住宅进入设计装饰的后期整理阶段时，负责该设计的年轻巴西设计师才与业主会面。这对夫妇对室内环境的布置和传统的装饰风格表示不满，所以他们全权委托设计师对公寓进行全面的重新设计和布置。

装饰设计以聚合物水泥、薄荷与蓝色调的融合为基础，整面墙壁上覆盖着在巴西几乎没有商业价值的马尾松木料。

该设计的另一个亮点是入口处的结构柱显露在外，并与餐桌共同构成一对完美的组合。很重要的一点是：所有主要物件，包括桌子、餐具柜和沙发，都是由设计师亲自设计的。业主原本就是 Guilherme 设计作品的爱慕者，所以在整个室内装饰设计过程中，她坚持购买几件由 Guilherme 设计的作品。

业主和设计师的合作关系如此融洽，以至于女主人最终决定开设一家公司，在巴西和世界范围内独家代理并销售 Guilherme Torres 的设计作品。其他与众不同的饰品包括：Moroso 品牌的灰暗色贵妃沙发，虽然它原本是一件室外家具，可 Guilherme 决定把它放置在公寓内；Diesel 地毯和钢制凳腿木质椅子则来自 Driade。由 João Machado 创作的艺术作品摆放在灰暗色贵妃沙发的旁边，餐具橱上面摆放着 Eliane Prolik 的作品。主卧室与起居室中色彩缤纷的装饰风格完全相反。白色和黑色的配色被应用于家具、床上用品，甚至室内框架的颜色上。床头板上方摆放着 João Machado 创作的艺术作品和 Ari Beraldin 床上用品。厨房内选择了黑色漆面家具和 Corian and Smeg Home Appliances 厨房工作台面。由于厨房和起居室之间没有分隔两个空间的门，因此起居室的墙壁也延续了厨房墙壁的设计风格。

LOCATION
Budynek Sea Towers, Gdynia, Poland

海塔公寓

SEA TOWERS APARTMENT

DESIGNER Magdalena Konopka, Marcin Konopka
DESIGN COMPANY Ministerstwo Spraw We Wnetrzach
AREA 100 m^2
PHOTOGRAPHER Marcin Konopka

An investor had approached the designers'studio and asked them to design interiors of a luxurious weekend apartment with functionality of every-day home. The designers have received a task to design an apartment from its basics: the investor handed us an open plan of 100 m^2. Functional and aesthetic guidelines received let the designers design an interior of clear division on common and private space. The common part consists of a living room with kitchen annex. A spacious hall leads to a study/guest-room separated by sliding glass door. The private space is invisible: together with a little bathroom it's hidden behind the door- integrated with wall lining. A bedroom with wardrobe and a bathroom are parts of the private zone.

Economy of colors and the leitmotif of white is a suggestion of the investor and a result of our perception of this inspiring space, where the sea plays the main role. Thanks to big windows, the apartment becomes a space of integrated inside and outside views. In order to minimize objects in the space, the equipment had been hidden in the custom furniture—it holds a canal heater, that together with side surfaces creates a seat with a beautiful view. The furniture also holds AV system, subwoofer and a bio fireplace, on the perpendicular wall. Above a TV set hidden behind a black glass panel is situated. The graphics gives background to the dark monolith. Few pieces of orange furniture and décor are contrasting accents in the investor's favorite color.

After sunset, when Baltic becomes disturbingly dark, the apartment gains a new life. Thanks to carefully chosen light and system of intelligent steering, the designers can create light scenes with ease and pleasure. Although there are a lot of white surfaces and lustrous textures, the apartment is pleasantly warm and welcoming.

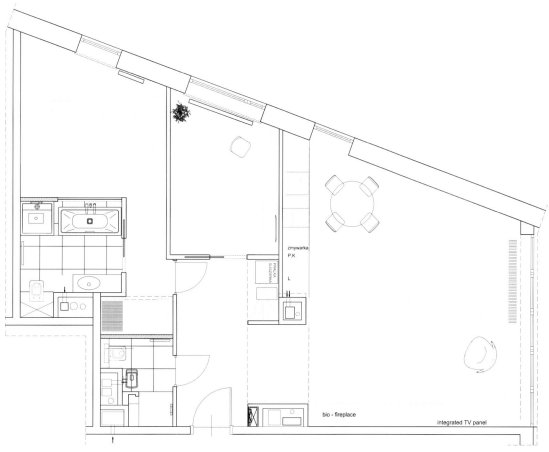

MSWW_ Apartment in Gdynia, Sea Towers

　　一位投资者来到设计师的工作室，请设计师对一间奢华的周末度假公寓进行使其拥有日常家居功能的室内设计。该设计的任务是从最基本的室内设计元素开始：投资人交给设计师一套面积为 100 平方米的开放式公寓设计方案。根据设计指导方针中的室内功能和审美要求，设计师精确地区分了公寓中的公共区和私密区。公共区包括一个起居室及其附属厨房。宽敞的门厅通向玻璃材质的拉门和与其隔离的书房（客房）。私密区并不一目了然：与一间小浴室共同隐藏在门的后面，并与墙衬相结合。一间配有大衣柜的卧室和浴室也属于私密区。

　　设计师综合考虑了投资者的建议，并且以其对休息区的理解和感知，在这个以大海为主旋律的空间内，采用了简洁的配色方案并选择了白色作为主色调。得益于明亮的大窗户，公寓内外的景色得以完美融合。为了尽可能多地减少空间内的物品，家用设备被隐藏在定制家具中。家具中容纳了管状加热器，并与其侧表面构成一个可以观赏风景的座位。它们还容纳了影音器材、次重低扬声器和垂直墙壁上的生物乙醇壁炉。在电视机上方的背面隐藏着一个黑色玻璃嵌板，平面造型为这一整块深色的庞然大物增添了一个背景。室内还摆放了一些色彩对比鲜明的橙色家具和装饰物，在投资者最喜爱的颜色中得到了凸显。

　　每当日落后，波罗的海的海面陷入一片令人不安的黑暗之中，这时公寓中开始了一种有别于这漆黑的夜的生活。得益于精心选择的照明设施及其智能操纵系统，主人能够轻松、愉悦地欣赏不同的照明场景。虽然大部分空间表面都选择了白色的配色和有光泽的表面纹理，整间公寓仍然洋溢着令人愉悦的温馨、热情的气息。

异国风情
EXOTIC &
CHARMING

LOCATION
Houston, TX, USA

SOUTHAMPTON MOROCCAN

南安普敦摩洛哥风格住宅

DESIGNER *Laura Umansky*
DESIGN COMPANY Laura U, Inc.
AREA 276 m²
PHOTOGRAPHER Julie Soefer

The designers created a exuberant and collected Moroccan style home for this well-traveled family. The couple wanted to be surrounded by their favorite pieces from their time spent in Morocco.

The designers renovated the home from top to bottom to create a Moroccan interior. Many areas have custom millwork (where budget allowed) and we created special nooks in the dining room and 3rd floor landing (part of master suite). The designers painted the exterior of the home, designed a custom entry gate, and also painted all of the interior doors vibrant cobalt. There were many custom touches in the master bath (not pictured). Also, the homeowners were both very tall, so many of the doors were raised to accommodate them.

　　设计师为这对游历甚广的夫妇打造出一个充满活力且舒适、惬意的摩洛哥风格的寓所。这对夫妇想要在室内摆放他们从摩洛哥带回来的充满他们在那里的美好回忆的物品。

　　设计师对整间公寓进行了彻底的更新和改造，以营造摩洛哥风格的家居环境。设计师在很多区域中放置了定制的风车（在装饰费用的预算之内）并且在餐厅和三楼的楼梯平台（属于主卧室套房中的空间）之间打造出一个特别的角落。设计师对住宅外部进行了粉刷，设计了一个专门定制的入口大门，并把空间中的门都涂成了钴蓝色。主浴室中的很多装饰也都是定制的（而不是绘制的）。因为这对夫妇身材都很高大，所以在设计过程中，设计师对空间中的很多门进行了加高处理，以满足他们的需要。

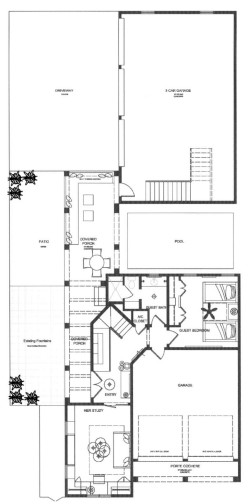
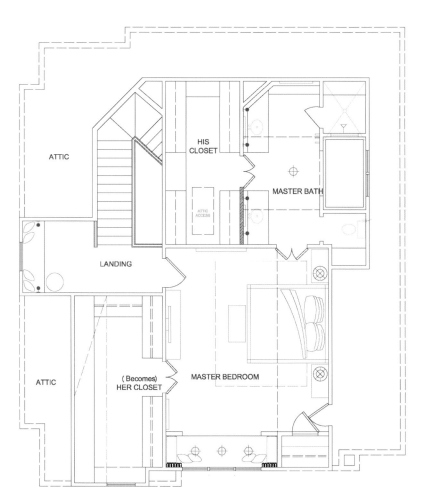

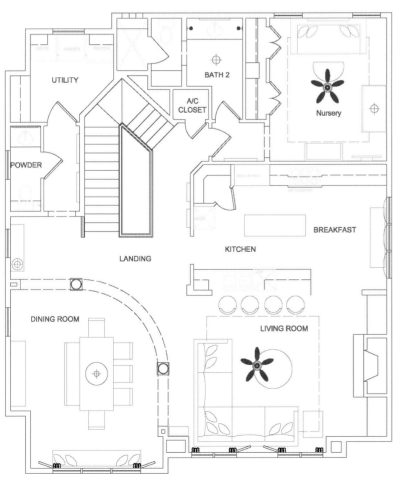
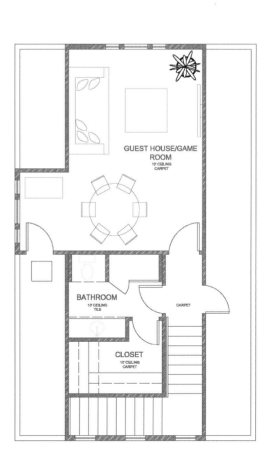

LOCATION
Ahmedabad, India

LUXURY AT EASE

DESIGNER *Hiren Patel*
DESIGN COMPANY Hiren Patel Architects
AREA 143 m^2
PHOTOGRAPHER Sebastian Zachariah

安逸舒适中的奢华享受

City homes are notorious for their lack of space and are mostly up to the architect to make the most of what's available. The designers have been instrumental in shaping the interiors on a plot measuring 1,115 m^2. The designers took up the daunting task of designing the interiors of this 743 m^2 bungalow and the surrounding landscape in the city where hot summers and arid winters are constant reminders of the desert that is always trying to reclaim it.

The result is a subtle, yet certain triumph of creativity, which has set an example for the other homes that are part of this housing scheme. The designers retained the basic structure of the house, which had to remain in uniformity with the other houses according to the gated community housing rules. The designers got involved early in the construction, so they could contribute to many aspects, like the flooring and the window design. The family room is the focal point of the house and seems to merge into the outer landscape, making it an ideal space for the family to spend time together. Since it is the core of the family interaction, this larger space was chosen to be the family room while the formal living room has been allotted a smaller share.

This space is equipped with a home theater and a serving counter. The furniture can be moved aside to double up as the dance floor. While the design of the upper areas of the house is sober and muted, Patel has given this space a more dramatic effect by combining various materials and color combinations to create a lounge like ambience. The light design adds the finishing touch of drama to all the elements in this space. The brass counter is encased in glass and matches the brass meshwork (jali) on the display cabinets. The overall effect is one of richness in simplicity. On the ground floor, one of the bedrooms has a wooden deck, which seems to float over the expansive lawn and acts as a private garden. The lawn itself is a pivotal part of the entire property.

Furniture Layout_Ground Floor

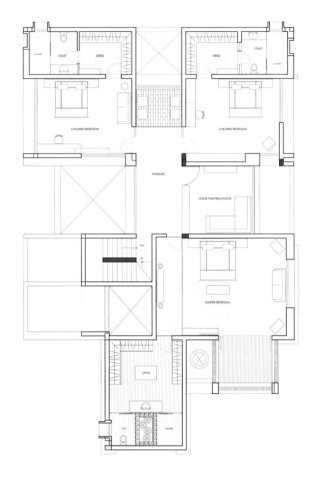

Furniture Layout_First Floor

  城市中的住宅以空间狭小著称，所以要依靠设计师去充分发掘和利用室内的空间。设计师曾经成功地完成了占地1115平方米的室内设计。所以这一次设计师接受了这个艰巨的设计任务，对这个室内面积为743平方米的夏季炎热、冬季干燥的独立住宅及其周边的室外环境进行设计。

  该设计是创造力的体现，也可以将其作为成功的样本，为今后同类的住宅建筑方案提供参考。根据该封闭小区的住房规定，为了与小区内的其他房屋保持一致，设计师保留了房屋的基本结构和框架。房屋地面、窗户等许多方面的设计都融入了设计师的设计理念。家庭室是整体设计中的重中之重，而且似乎逐渐与室外的自然景观融为一体，是理想的家庭聚会场所。因家庭室为家庭成员互动交流的主要场所，故设计师选择了住宅中面积较大的房间作为家庭室，而住宅中正式的起居室则被安排在一个面积相对较小的房间内。

  家庭室中包括家庭影院和服务台。可以把室内的家具移开并集中摆放，腾出原有双倍的空间作为舞池。住宅楼上的区域选择了较严肃和柔和的装饰风格。通过搭配使用不同的色彩和装饰材料，设计师赋予了这个和周围环境气质相符的休息室更加引人注目的风格。灯光设计是该设计的点睛之笔。玻璃表面的黄铜材质柜台和展示橱柜中的黄铜网状艺术品相搭配。室内空间的整体装饰效果是于简单中凸显丰富和华美。住宅地面层的一间卧室中铺设了一个木质平台，使整个房屋好像漂浮在室外广阔的草坪之上，如同一个私人花园。而室外的草坪也是这座花园住宅的关键组成部分。

LOCATION
Ahmedabad, Gujarat, India
THE FACILITATION

DESIGNER *Hiren Patel*
DESIGN COMPANY Hiren Patel Architects
AREA 282 m²
PHOTOGRAPHER Sebastian Zachariah

简易化公寓

When the client has faith to only see the final finished interiors, it speaks volumes of the designers' abilities on one hand and on the other, challenges them to meet his lofty expectations. The designers aimed for minimalist, contemporary sophistication without compromising on ease and comfort in this home away from home.

Less is more
Minimalism means space openness. So, one bedroom was deducted from this four-bedroom apartment. An open plan living room was created adjoining the guest bedroom, and one master bedroom was fashioned from two earlier bedrooms. Further, the apartment was widened to accommodate a verandah, so the home could breathe and share light.

Color thrift
White has been retained as the color of most walls to accentuate the client's art collection. But, small splashes of color have been used to bind the spaces together—an olive green wall connects the lounge to the dining space and bedroom, a plum wall adds a note of rich comfort to the living area, while a dark chocolate wall background makes the leaf mirrors here stand out.

Luminous art
The lighting used is stylish designer lighting. The ceilings have very simple LED lights, but in the couch area, a play of hidden LED light within the gypsum ceiling brilliantly showcases the painting and object d'arts. Wall fittings have been generally avoided, and lighting beside the beds, floor lamps, reading lamps, etc, have been treated like art objects.

客户只看最终的装饰效果,一方面,这可以充分展示设计师的能力,另一方面,也是带给设计师的一个巨大挑战,看他最终的设计成果能否达到客户的期望。该设计以极简主义设计风格为目标,既彰显出现代装饰风格的精致,又使这个远离故土的居所不失安逸和舒适之感。

"少即是多"的设计理念

极简主义意味着空间开放性。所以,设计师移除了公寓中四间卧室中的其中一间。一个敞开式的起居室与客卧室相连,主卧室也是由原来的两间卧室合并而成的。并且,整间公寓被扩宽,以容纳一个外廊,使人能够呼吸室外的新鲜空气并接受室外的光照。

色彩节俭主义

室内空间中的大多数墙壁上使用了白色,以凸显房屋主人摆放的艺术品。但在装饰过程中也广泛运用了小块的色彩点缀,使各种元素相融合。休息室和用餐区、卧室之间的墙壁选用了橄榄绿的配色,生活区内的紫红色墙壁对整个空间的舒适感进行了诠释,深巧克力色的背景墙凸显出摆放在室内的叶形镜子。

光照的艺术

室内空间采用了现代风格的灯光设计。天花板上安装了简易的LED照明。但是在摆放沙发的区域,其上方的石膏天花板上安装的隐藏式LED照明鲜明地凸显出装饰油画和陈列的艺术品。在整体空间中,避免了壁灯的使用,床边的照明灯、落地灯、台灯等灯饰都如同艺术品一样在空间内进行了合理的安置。

Furniture Layout

LEGEND
1 - Entrance Foyer
2 - Vestibule
2A - Puja
3 - Living Room
4 - Library
5 - Family Space
6 - Lounge
7 - Dining Room
8 - Kitchen
8A - Store
8B - Wash
9 - Bedroom
10 - Toilet
11 - Guest Bedroom
12 - Servants Room

LOCATION
Ahmedabad, Gujarat, India

THE AMALGAMATION

DESIGNER *Hiren Patel*
DESIGN COMPANY Hiren Patel Architects
AREA   690 m²
PHOTOGRAPHER Sebastian Zachariah

The backyard of the old home has been transformed into a private garden with a meticulously planned layout. Each plant has been handpicked and nurtured. A stone-carved waterfall with flower and butterfly motifs amongst greenery along with mood lighting makes this a perfect party and entertainment area. It is simple in its design and yet spectacular with its glass walls and specially fabricated silver-plated carved door. The six panels on the door are sourced from Udaipur and framed locally. The temple was shifted from the old house to the outside, according to Vastu.

The patio is shaded by a wooden awning, which stands on a six antique pillars sourced from a dealer on the outskirts of the city. Foliage, stone and terracotta statues and carvings and a main door of teak wood with intricate paneling, welcome one in. The formal living areas is an amalgamation of the old and the new, some pieces inherited, some bought. Pichwais, statues, sculptures, silverware and artifacts sourced from various trips abroad, mainly Thailand and China, exist in perfect harmony. The two identical wall units are family heirlooms. One side of the living room leads to the dining room, which further extends into an outside area.

A metal buffalo greets you, when you walk through the door leading to the informal living area, as a beautiful glass lampshade redesigned from an old chandelier. The traditional Jhoola(swing) is the focal point here. It was custom-made by a local carpenter, like the rest of the furniture. The upholstery is in a riot of bright colors, which animate the ambience. Upstairs a common room leads to three bedrooms with attached bathrooms. The couple's two sons share a bedroom and a study. The bedrooms are specifically designed to be functional and elegant, yet understand-unlike the rest of the house.

住宅原来的后院被改造成精心布置的私家花园。花园中的每一株植物都是精心挑选并精心培育的。坐落在绿色植被中带有蝴蝶和花朵图案的石刻人工喷泉,在情景灯光的搭配下,使整个花园成为一个聚会和娱乐的绝佳场所。整座花园的设计虽然简单,但配置的玻璃墙和采用特殊工艺焊接的镀银雕刻门使整座花园的景色引人入胜。板式门上的6块门板的材料来自乌代布尔,并在本地进行装框。与此同时,设计师把住宅后院中的供堂搬到了外面。

露台中配置了一个木制遮阳棚,是从市郊的一个商人那里购买的,由6根古式柱子作为支撑。叶饰、石头、赤土陶雕像、雕刻品和表面花纹错综复杂的柚木材质镶板正门,迎接着来访者。正式的生活区融合了新旧装饰,既保留了一些原有的物件,又包含了一些新买的物品。手工编织的装饰画、雕像、雕塑、银器和手工艺品搭配和谐,大都是去泰国和中国的旅行时带回来的。两个相同的组合柜是家里的祖传遗物。起居室的一边通向餐厅,一直延伸到外部区域。

穿过大门,进入非正式的客厅,首先映入眼帘的是一个金属材质水牛饰品和由一个古旧的吊灯经过重新设计而来的一个漂亮的玻璃灯罩。传统样式秋千

The Amalgamation_Site Plan

是室内设计的点睛之笔。和其他家具一样，这个秋千也是在当地的一位木匠那里定做的。家具的垫衬物的面料也是五颜六色的，使整个空间充满活力。二楼的公共休息室通向三间带有室内卫生间的卧室。业主的两个儿子共享一间卧室和一间书房。卧室设计考虑到了空间功能性，营造出优雅的空间气氛，与其他房间不同，散发着一种知性魅力。

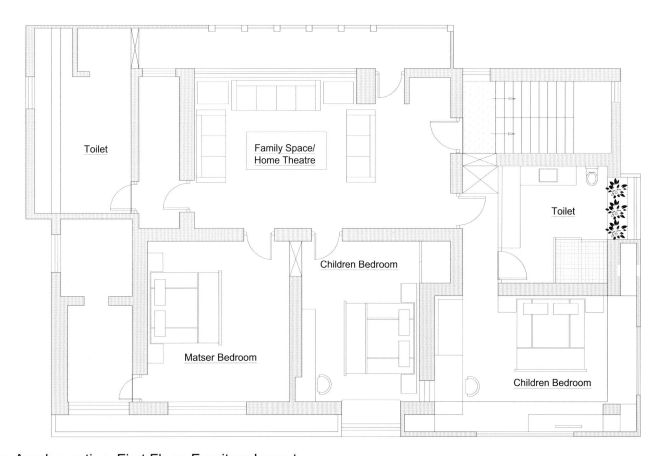

The Amalgamation_First Floor_Furniture Layout

LOCATION
Park Slope, Brooklyn, New York, USA

Park Slope 住宅"复兴"

PARK SLOPE REVIVAL

INTERIOR DESIGN Erin Fearins, Ward Welch, Catherine Brophy

ARCHITECT   Brendan Coburn
FIRM   CWB Architects
TEAM   Janelle Gunther, Joseph Smith, Romina Romano
AREA   348 m^2
PHOTOGRAPHER   Hulya Kolabas

This mid-19th century row house in Park Slope was in relatively poor condition when purchased by the client. The house next door had burned down many years before, leaving the lot vacant and overgrown, and the new owners wanted to build a structure in its place that would connect to the existing house.

With the client's ambitious program for the new structure and to avoid having to move to temporary quarters, the client elected to complete the project in two architectural phases, followed by an interior design phase. The double parlor was designed in one dominant theme to give it the feel of a single space. As a family of four, the client had a need for substantial storage space. Extensive millwork accommodated books, music and electronics. Because the client opted to locate the master bedroom on the parlor floor, special care was given to the delineation of the space as private quarters. The vestibule library, with its extensive built-ins, served as a threshold to the master suite. The dining room was completed during the second architectural phase; it adjoined the kitchen. A spacious and functional kitchen opened onto the patio garden.

GARDEN FLOOR

1 Entry
2 Living Room
3 Powder Room
4 Kitchen
5 Dining Room
6 Garage

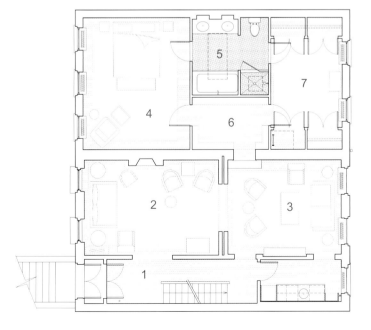

PARLOR FLOOR

1 Foyer
2 Living Room
3 Family Room
4 Bedroom
5 Bath
6 Library
7 Dressing Room

这座位于 Park Slope、建于 19 世纪中叶的连栋住宅在住户购买之初整体状况不佳。住宅的隔壁房屋在多年以前就已经被烧毁，使房屋周围遗留出大量空地，杂草丛生。而房屋的新主人则想要在这片空地上建造一个建筑，与原有的房屋相连。

由于客户雄心勃勃地要建造这个新建筑，并想要避免搬到临时住所，因此他选择在完成房屋的室内装饰设计之后，分两个建筑阶段去完成这一项目。室内的两个会客厅是在一个支配性的设计主题的指导下完成的，给人一种单一空间的感觉。由于家中有四个家庭成员，因此房屋内需要大容量的储存空间。广泛使用的木质建筑材料与摆放的书籍、音像制品和电子产品相搭配。由于客户想要在客厅层的空间内打造主卧室，因此设计师特别注意了其作为私人生活区的整体轮廓。前厅图书室中加入了大面积的嵌入式结构，作为进入主套房的入口。餐厅的设计完全处于室内设计的第二阶段，并与厨房相邻。功能齐全、空间宽敞的厨房向住宅的内院敞开。

LOCATION
Vero Beach, Florida, USA

WINTER HOME IN VERO BEACH

Vero 海滩的冬日之家

DESIGN COMPANY Robert Couturier & Associates
AREA 557 m²
PHOTOGRAPHER William Waldron

Entrance Hall:

Antique "Tuareg" straw and leather rug

Anglo Indian round table

Red lacquer Chinese cabinet, 18th Century

Living Room:

A very important Indian Dhurrie rug, 19th Century with a central medallion of a lion

A pair of Custom sofas designed by Robert Couturier covered in white linen

A pair of Jean Michel Frank style waterfall coffee tables in straw marquetry

A custom square ottoman with wooden base designed by Robert Couturier

An Anglo Indian swing, 19th Century

Walls–custom painted

Curtains in 100% linen from Bruder, Paris

Family Room:

Custom sectional sofa designed by Robert Couturier in yellow Fortuny fabric

Anglo Indian armchairs

Green curtains in 100% linen from Bruder, Paris

Dining Room:

Walls–covered in "IKAT" panels

Rug in linen and cotton by A.M.Collection

Anglo Indian table and chairs

Anglo Indian chandelier

White linen curtains from K-5 Linen

Kitchen:

All custom cabinetry

Playroom:

Custom "Map" wallpaper

Daybed designed by Robert Couturier in blue & yellow linen

Bean bag from pottery barn

Her Office:

Wall upholstery in "Raoul" Textiles fabric

Armchairs by Holly Hunt

Boy's Room:

Beds–custom made

Her Bathroom:

All work–custom made

All woodwork–custom made

Custom hardware by Carl Martinez

Curtains in white linen with pink border by Bruder, Paris

Guest Room:

Anglo Indian bed, 19th century, India

Lamps in orange suede, custom made

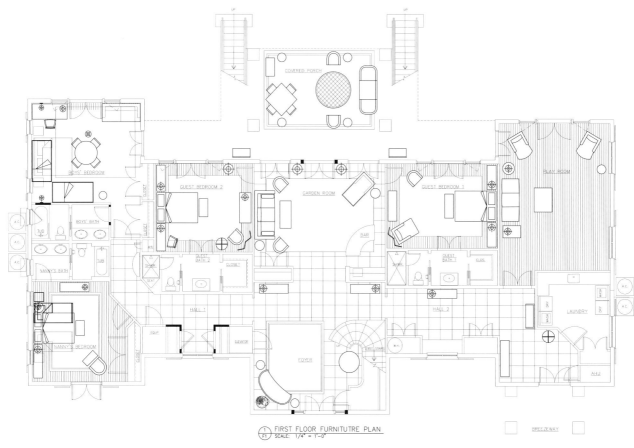

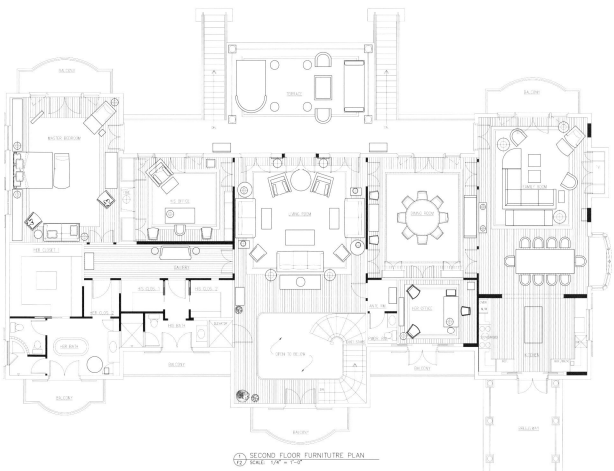

SECOND FLOOR FURNITURE PLAN
SCALE: 1/4" = 1'-0"

门厅：

古式塔雷格稻草和皮革编织的地毯。

英印风格圆桌。

18 世纪的红色喷漆中式柜。

起居室：

重要的装饰元素：一块 19 世纪的印度手纺纱棉毯，棉毯中央是一个狮子奖章图案。

一对由 Robert Couturier 设计的白色亚麻面料定制沙发。

一对 Jean Michel Franks 风格的稻草纤维镶嵌而成的瀑布式咖啡桌。

由 Robert Couturier 设计的有木质底座的方形搁脚凳。

19 世纪的英印风格秋千。

定制喷漆的墙壁。

来自巴黎布吕代的百分百纯亚麻面料窗帘。

家庭房：

由 Robert Couturier 设计的黄色 Fortuny 面料定制组合式沙发。

英印风格扶手椅。

来自巴黎布吕代的百分百纯亚麻面料绿色窗帘。

餐厅：

宜家品牌的面板。

由 A.M.Collection 提供的棉麻材质地毯。

英印风格桌子和椅子。

英印风格吊灯。

K-5 亚麻面料白色窗帘。

厨房：

定制橱柜。

游戏室：

定制"地图"墙纸。

由 Robert Couturier 设计的蓝色和黄色亚麻布料坐卧两用长椅。

购自 Pottery Barn 的懒人沙发。

办公室：

Raoul 纺织装饰面料。

Holly Hunt 扶手椅。

男孩房：

定制的床。

浴室：

所有物件都是定制的。

所有木制品都是定制的。

Carl Martinez 定制五金器具。

来自巴黎布吕代的白色百分百纯亚麻面料粉色镶边窗帘。

客房：

印度 19 世纪的英印风格的床。

橙色仿麂皮面料定制灯具。

LOCATION
Kohala Coast, Hawaii, USA

KUKIO RESIDENCE

DESIGNER Jacques Saint Dizier, ASID
DESIGN COMPANY   Saint Dizier Design
ARCHITECT   Blanca Geesey, Peter A Geesey, AIA
AREA   743 m^2
PHOTOGRAPHER   Mary E. Nichols

The house itself is located on the Kohala Coast of the island of Hawaii. It has a Main hale, or pavilion, and two Guest hales for a total of about 743 m^2.

The architects Blanka and Peter Geesey created a home that blurred the lines between inside and out. With our interiors, we tried to maintain the feel of Polynesia, but with a bit of a glamorous edge. For instance, the Master Bedroom walls—floor to ceiling—are covered in mother of pearl. The bath tub is a beautifully painted copper set in black ili-ili stones. There are lush gardens surrounding the house and all through the grounds with meditation benches and sculpture in various spots.

All of the art was curated by the designer and is all either by a Hawaiian artist or connected to Hawaii in some way. The pool features a swim-up bar with a thatched roof that is done in blue mosaic tile. From the lanai, one can watch the beautiful Hawaiian sunsets and enjoy the tropical island breeze.

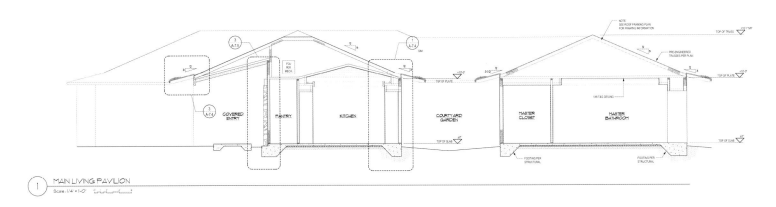

① MAIN LIVING PAVILION
Scale: 1/4" = 1'-0"

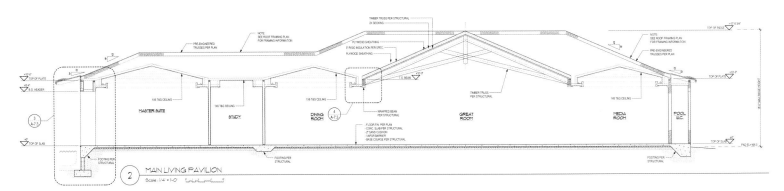

② MAIN LIVING PAVILION
Scale: 1/4" = 1'-0"

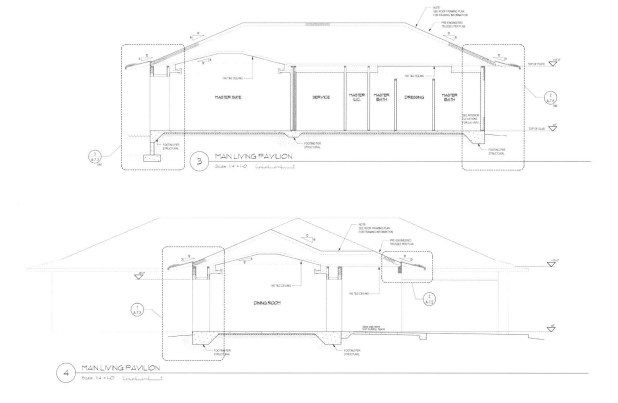

OVERALL SITE PLAN

房屋位于夏威夷岛的柯哈拉海岸上。室内设有一个主阁间或阁楼以及两个客用阁间，总面积大概为743平方米。

设计师 Blanka 和 Peter Geesey 在打造室内空间的过程中，有意模糊了室内与室外的线条。设计师试着在使室内保持玻利尼西亚岛的风格，并融入了一些迷人的边线设计。例如，主卧室的墙壁，从地板延伸至天花板，覆盖着珍珠母。室内摆放着嵌在黑色的伊犁石中泛着迷人的金属光泽的铜制浴缸，房屋四周环绕着植被苍翠繁茂的花园，花园中随处可见用于冥想静思的长椅和摆放在不同位置的雕塑。

住宅内摆放的所有艺术品都是由设计师精心挑选的，它们或者出自一位夏威夷的艺术家之手，或者与夏威夷有着某种联系。游泳池的设计亮点是与其搭配的带有蓝色马赛克瓷砖镶嵌的茅草屋顶的池滨酒吧。在阳台上，可以欣赏美丽的夏威夷日落并尽情享受这个热带岛屿上的阵阵和风。

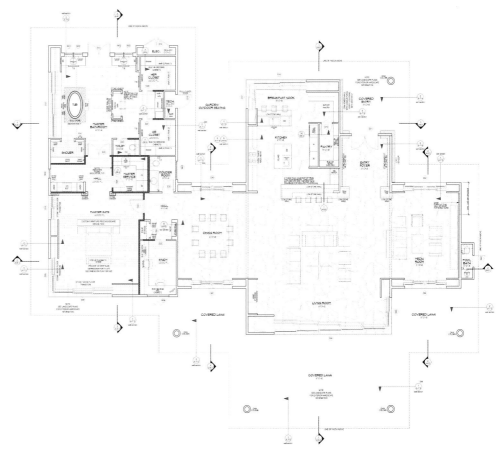

INDEX 索引

Laura Umansky

Laura Umansky is an interior designer who brings her signature aesthetic to luxury international residences, fashion-forward hospitality interiors, and high-style corporate offices. Her classic architectural education and international design experience is channeled into each and every design decision. At Laura U, Inc. she leads a talented team focusing on turn-key interior design solutions. Over the last five years, Laura U, Inc. has won 12 ASID awards. Laura Umansky has been recognized by the interior design industry through additional awards: ASID Presidential Citation, Houston Design Center's Inaugural Stars on the Rise, Houston Business Journal's Top 40 Under 40, and Luxe magazine's National Gold List. Laura received her Master of Architecture degree from the University of Houston and her BA from the University of Texas (Austin). She is an Allied Member of ASID and an Associate Member of AIA. Currently, Laura is designing luxury residences throughout the US and bringing new hospitality concepts to life.

Karim Rashid

Karim Rashid is one of the most prolific designers of his generation. Over 3000 designs in production, over 300 awards and working in over 35 countries attest to Karim's legend of design. His award winning designs include democratic objects such as the ubiquitous Garbo waste can and Oh Chair for Umbra, interiors such as the Morimoto restaurant, Philadelphia and Semiramis hotel, Athens. Karim collaborated with clients to create democratic design for Method and Dirt Devil, furniture for Artemide and Magis, brand identity for Citibank and Hyundai, high tech products for LaCie and Samsung, and luxurious goods for Veuve Clicquot and Swarovski, to name a few. Karim's work is featured in 20 permanent collections and he exhibits art in galleries world wide. Karim is a perennial winner of the Red Dot award, Chicago Athenaeum Good Design award, I.D. Magazine Annual Design Review, IDSA Industrial Design Excellence Award.

Ministerstwo Spraw We Wnetrzach

MSWW - Ministerstwo Spraw We Wnetrzach (in easy translation—Ministry of Interior Affairs) is a interior desing studio founded by Magdalena Konopka and Marcin Konopka.
MSWW works mainly in the area of the Tri-City in Poland (Gdansk, Sopot, Gdynia). Nevertheless, also takes actions located away from our region—also abroad. They devote their energy and engagement into smuggling beauty to the space of everyday life. Their aim is to shape places where living, eating, working and leisure have their own, special and individual dimension. They see interiors as a mosaique of elements that try to arrange into a harmonious entirety. To gain proper effect MSWW deal with designing furniture, lighting, crafts as well as other activities that fit into the space or stay on the fringe of design. The area of MSWW activities includes interior design seen from a wide perspective. MSWW design is mainly a contemporary style with a touch of old times and class. It has also a wise look to the future, modernity, and futurism. Those things create a conglomerate of style filled with a huge amount of emotions.

Robert Couturier

Couturier is known for creating visions that relate to historical moments, infused with the presence of today. His projects, whether Mexican estates or Manhattan apartments, are lush, light, sensual, experimental, and witty. Regardless of how aesthetically beautiful or freshly inventive his creations are, he insists on comfort. Hallmark's of Couturier's work include the particular concern for the ways people move, live, and function in the spaces he designs, and the ability to draw inspiration from a variety of periods that span from the 18th century to the modernists of the 1940s, including such icons as Renzo Mongiardino, Frank Gehry, Charles Le Brun, Poillerat, Serge Roche, Robjohn Gibbons, Mallet-Stevens and Jean Michel Frank. Couturier's prodigious talent propelled the Paris-born designer into Architectural Digest's top 100 Designers in 2000, 2002, 2007, and in January 2012 he was once again recognized on this coveted list. In June 2012, he was included on Elle Décor's A-List Top 60 Designers as well as top 10 foreign decorators in British House & Garden. He has also been acknowledged by the Franklin Report as "innovative, creative, and stylish" in his interior design.

Jamie Bush & Co.

Widely admired for his relevant and keen understanding of architecture and design, Jamie approaches each new commission with a discerning and insightful point of view. Growing up in a family of renowned designers, photographers and artists in New York propelled him to study architecture and design in New Orleans and in Italy. After receiving his Masters in Architecture from Tulane University, Jamie headed west to practice architecture in Los Angeles and founded his own interior design firm in 2002. He is recognized for his ability to mix period and contemporary furnishings with a fresh and discriminating eye. Layering rich colors and textures with exotic and organic elements transforms each of his spaces into an inviting, modern and unexpected environment. Jamie's highly regarded designs have been featured in over 40 publications world wide including Architectural Digest, Interior Design, Casa Vogue, Dwell, Los Angeles Times as well as several coffee table books.

Jennifer Dyer

Jennifer graduated from the Univ. of TN Knoxville with a double major in psychology and sociology all the while knowing that her interest truly lay in the field of interior design. After graduating she was offered a position abroad and resided in London for 4 years. Returning to the United States on the brink of the millennium, she took a risk and moved to Miami, FL to pursue a career in design. One of Miami's top design firms offered her full time employment after only a 6 month internship. She began working closely with a team of multi-skilled designers on extremely high end projects through-out Latin America, Europe and the United States to learn the trade. Due to the design firms' popularity and the real estate boom of the early 2000's, she was able to design and project manage dozens of full scale residential and hospitality design jobs. Her employer also operated an exclusive retail showroom of which she became one of three retail buyers at trade shows in New York City, Dallas, Paris, and Milan.

Elias Kababie

Elias Kababie graduated from the Faculty of Architecture at Universidad Iberoamericana in Mexico City. He continued his studies at Parsons Design School in New York specializing in industrial and interior design and complemented them with lighting courses at Phillips and architecture courses at Pratt Institute.His professional career is a balance between architecture, interior and product design. His main goal is to achieve perfect synchrony between architecture, interiors and objects in order to transmit the lifestyle of each one of his clients. Under the motto "back to basics", projects created by Elías are proposals beyond the creative boundaries with a unique concept that includes technology, current trends and global expression. His designs have been published in important magazines and newspapers in Mexico and in other countries and have also received important awards. He has participated in major stores like CAD, Common People and Casa Palacio and some of his designs are being sold at the Mexican Gallery of Design and in MUAC and MODO museums in Mexico City.

Hiren Patel Architects

In my architectural practice spanning a rich 25 years, I have gone through several ups and downs, small ones and big ones...growing with every challenge ...adapting to prevailing conditions to tide me through the worst crisis. This journey has taught me through wonderful experience to introspect about the intricate idiom with ease. In my quest for excellence, I have poured into my work passion, commitment, whole- hearted enthusiasm, hope and my energy. And I hope to maintain just such a level of integrity and commitment in all my work as long as I live. Hiren Patel architects started out as a small firm that has grown exponentially by accepting every challenge that came its way. Our initial success came from designing high rises in Ahmadabad, our base. Thereafter, we have not looked back by aiming ever higher. Today, our palette of work covers everything from a small residence to heritage buildings and palaces to huge entertainment complexes, from an individual shop to large commercial megaliths.

Maurício Arruda

Brazilian architect and designer Maurício Arruda develops and executes projects in architecture, scenography and product design in his studio Maurício Arruda Arquitetos + Designers. Master in Architecture, specialized in Sustainable Construction, Maurício took part in the development of Agenda 21, which sought to guide new perspectives in Sustainable Building to developing countries. His design creations full of color, humor and brasilianity also carry environmental and social concerns, result of an extensive research on the fields. Maurício is also a design professor since 2001 and have lectured in several institutions, having IED - Istituto Europeo di Design, Escola Panamericana de Arte and SENAC among them. In contemporary architecture and design, Maurício is known as a Brazilian reference in such fields, being invited to showcase his work in galleries such as the Flanders Gallery, in Brussels, Belgium and The Gallery at the civic, in Barnsley, England.

Casa Forma

Casa Forma is an architectural and interior design firm comprised of a team of 13 architects/interior designers headquartered in London. Their experienced team is led by world-class experts from the fields of decorative and structural design. Their international practice provides a comprehensive service providing clients architectural and interior design services for luxurious residential, commercial or hospitality properties. For Casa Forma's CEO Faiza Seth, the desire to improve the way that people live lies at the root of all Casa Forma work. They believe in providing truly bespoke design that is designed specifically for each client. Casa Forma's mantra is "functionality", making each space work as hard as it can to meet one's needs and increase the value of a property at the same time. After the functionality of a property is maximized, Casa Forma's architectural and interior design team designs the interiors based on a clients' personal style, and lifestyle.

Elizabeth Gordon

A designer with a professional background in high-end commercial projects and other entertainment venues, Elizabeth's creative foundation is reinforced with both a bachelor's degree in interior design as well as a Master's in Architecture. She has spent the last twenty years on projects within the hospitality and entertainment industries, refining her experience designing and project managing high-end commercial and international commissions including destination-caliber hotel resorts, spas, restaurants and retail stores. Elizabeth Gordon has now focused her talents and experience on projects in her own signature style including exclusive retail and office spaces in Beverly Hills, and high-end residences in Seattle, Los Angeles and the Hollywood Hills.

Pepe Calderin Design

Pepe Calderin Design is a Miami-based interior design and architecture firm with over 20 years of experience in high end residential and commercial design. Specializing in innovative and modern design throughout the Americas, Europe and Saudi Arabia, services include complete design concepts, space planning, construction documents, permitting plans, 3D renderings, specification and purchasing of materials, furniture and lighting, and installation. As the recipient of numerous national and international design awards, Pepe Calderin Design has garnered recognition for its fresh, energetic visionary approach to the design process, where the philosophy is "A space has no boundaries, and has endless possibilities."

Jacques Saint Dizier

Jacques Saint Dizier is the President of Saint Dizier Design Associates in Healdsburg, California and owner of the boutique home furnishings showroom Saint Dizier Home. The firm has won numerous awards for their work and Mr. Saint Dizier has twice been listed by Architectural Digest magazine as one of the top 100 designers in the world. Mr. Saint Dizier has served on the Board of Directors of the American Society of Interior Designers in New York City where he steered the Events Committee and was Chairman of the Industry Relations Committee. He is involved with many non-profit organizations including Face-to-Face, the Sonoma AIDS network, Food For Thought, the Healdsburg Food Pantry and many others. He is President of the Board of Directors of Re=Action Foundation, a non-profit organizations dedicated to emerency disaster relief. A graduate of Louisiana State University, Mr. Saint Dizier started his career in New York City where he practiced for 12 years before relocating to Northern California.

Jessica Lagrange

Jessica Lagrange brings over 30 years of experience to her eponymous interior design firm. Jessica Lagrange Interiors (JLI) provides its clients with a full range of interior architecture and design services, including color, material, and furniture selection, product specification, and oversight of construction and installation in both high-end residential and commercial markets. With services that extend to all aspects of the interiors, her work reflects a luxurious lifestyle, blending clean architectural lines, classic furnishings, and subdued color and texture. Jessica began her career at the Chicago office of Skidmore, Owings and Merrill and as an interior designer at the architecture firm of Powell Kleinschmidt. She earned her Bachelor of Fine Arts from Chicago's School of the Art Institute's Department of Interior Architecture.

Pebbledesign / Çakıltaşları Mimarlık Tasarım

Pebbledesign, founded in 2004 in Istanbul, especially focused themselves for designing of residential and commercial purposes as apartments & residentials, sample flats, concept & decoration projects, restaurants & cafes, exhibition booth, office and yachts. Founder of the Pebble Design, Interior Designer Ms. Neslihan Pekcan born in 1976 in Istanbul, after attending Savannah College of Art and Design "Interior Design Department" / USA, graduated from Mimar Sinan University "Architecture Faculty/Interior Design Department" in 1999. After her graduation, she started and continued her professional career in various architectural companies. Pebbledesign produces simple, absolute and particular solutions that the details form their work. Nonetheless, they autograph to the boutique projects by preferring reflecting their design pleasure, due to answering all needs of their clients.

Mariangel Coghlan

Interior Designer, passionate creator of designs, columnist and lecturer, with a clear vision of the social and humanistic responsibility required by the profession, Mariangel Coghlan is a leader in Mexican interior design. She has won many awards for her projects and has extensive experience in creating spaces. She has her own interior design firm: MARIANGEL COGHLAN, where she leads a talented interdisciplinary team that provides excellent service and creates spaces that encourage people to gather, promoting a deeper understanding of the people with whom we interact. The guiding principle behind her designs is one she, herself, created: FUSION + MEXICO, which is the result of her reflection on the international interdependence of interior design trends and the incorporation of the wonderful shapes, colors and natural resources, so abundant in Mexico.

Fabio Galeazzo

Know for joining together beauty and sustainability, Fábio Galeazzo work with timelessness. Creative in everything he proposes, he mixes materials with colors in a feisty and fearless way. Being know for a professional that don't repeats himself, his work has soul, where the past and the comtemporary live together in total harmony. He plays with the diachrony completely free and this is one of the trademarks of his language. Reread the time in the objects and the objects of time with his own style. In 2004, founded Galeazzo Design, multidiciplinary company of architecture, interior design and product development, with a young and aligned team, being awarded nationally and internationally. His work is gaining prominence among the internacional press and it's been published in more than 50 countries.

Design Intervention i.d.

Design Intervention i.d. was established in Singapore in 2004. Our Design team comprises Interior architects, stylists and designers, project managers and architects. Thereby enabling us to take a project from conception to fruition; to create a totally comprehensive and coordinated look for the ultimate bespoke home, finished and tailored to the client's individual lifestyle in every way. Our team is truly international: between us, we have worked with clients spanning five continents; on projects as diverse as show flats in London and Dubai, hotels in Western Australia, New Orleans and London; as well as a vacation retreats in the French countryside, Swiss Alps, beach homes in Australia, Bali and Ski Lodges in Japan. Our team of 25 is made up of 11 different nationalities - each bringing their own unique experience to the Design Intervention team.

Rolf Ockert

Prior to graduation from the highly regarded University in Stuttgart, Germany, Rolf spent extended periods of time gaining practical experience in renowned architectural practices in Germany, England and Japan, Amongst them Frei Otto, James Stirling, Michael Hopkins and Riken Yamamoto.
Rolf Ockert Design, commenced in 2004, has in a relatively short period of time created a rich portfolio of work, ranging from product design to residential, commercial and retail projects and masterplanning, located throughout Australia and also overseas, most recently in New Zealand, Japan and Switzerland. Many projects have been published in national publications as well as in books and magazines in, amongst others, Italy, China, Israel, the US and Russia with others in line to be published soon. Rolf is a member of the Australian Institute of Architects and the Board of Architects in NSW and Victoria.

Kari Arendsen

Kari Arendsen is a world renound Interior Designer, honored by Andrew Martin Interior Design Review as being one of the top 100 Interior Designers Globally. Arendsen has been designing custom interiors for over ten years, with a specialization in new construction and remodels throughout the United States. Her works have been highlighted in San Diego Home and Garden, Ranch and Coast Magazine, Luxe Magazine, Premier Magazine as well as many other print and on-line media such as HGTV on line and "Kitchen of the Week" on Houzz.com. In addition, she has been the featured designer for the San Diego Historical Society's project at Liberty Station in Point Loma and Set Designer for "The Hungry Woman," a 2008 nationwide feature film. In 2013 she was an honorary Designer for the Pasadena Showcase House of Design. In addition to her thriving Interior Design Firm, Kari also has a Non Profit, Pillow Talk.

Guilherme Torres

Guilherme Torres can be defined by Daft Punk's misc: work it harder better faster make it over, the sentence has become his motto and it has been tattooed on his arm and written on the walls of his studio. His professional career started very early, when he was still a teenager, he worked as a designer in an engineering office, where he acquired the knowledge in the area and also the technical language. A world citizen, Guilherme divides his time between his studios in Sao Paulo and Londrina, with a range of activities, from residential and commercial projects to furniture design, one of his passions. Owner of his own and authorial style, his works were granted many awards and publications. For 2011, Studio Guilherme Torres is preparing to become an international studio, as result of the global media recognition.

Tommy Chambers

Los Angeles interior designer Tommy Chambers opened his own design company, Tommy Chambers Interiors Incorporated, in 2001 after eight years at Chambers and Murray, Inc. as head of the interiors department and the senior business and operations manager. Prior to starting Chambers and Murray, Inc. Tommy apprenticed for five years with the venerable interior designer, Joan Axelrod of Joan Axelrod Interiors. In addition to receiving an architecture degree from Texas A&M University, Tommy spent a year in Europe studying art, architecture and decorative arts. Tommy is listed in the (Los Angeles Franklin Resort), an independent evaluator of designers and design services. He is also an active professional member of the American Society of Interior Designers (ASID), the International Interior Design Association (IIDA) and the Institute of Classical Art and Architecture(ICAA).

Vernon Applegate & Gioi Tran

Applegate Tran founders and principal designers Vernon Applegate and Gioi Tran are both based in San Francisco, where the firm was founded in 1998. Vernon Applegate was educated at The University of the Arts, in Philadelphia, where he received a B.S. in Architectural Studies, and has done additional coursework through the University of California. He worked for several notable interiors design firms in Los Angeles and San Francisco prior to opening ATI. Gioi Tran received a B.A. in Interior Design from the Academy of Art College in San Francisco. He is recognized for his expertise in kitchen and bath design, and is frequently interviewed in industry publications. Gioi has served as a board member for Northern California Chapter of NKBA.

Boscolo

Boscolo has over ten years' experience in designing and delivering exquisite, luxury residences in some of London's finest locations. Creating unusual intimate apartments; finding, designing and preparing houses for those relocating from abroad; and remodelling and detailing substantial houses with gardens, terraces and conservatories. For some clients, we concentrate on the core decoration and furniture, for others we project manage a broader scope of work, including co-ordinating architecture and planning, remodelling, integrating entertainment systems, lighting, landscaping and installing leisure facilities of all types and sizes.

CWB Architects

When CWB (Coburn Welch Boutin) Architects was founded in 1995, it was a two-person practice focused on historic row house renovations. Since then, it has expanded into a multi-disciplinary firm which creates residential and commercial spaces throughout Manhattan, Brooklyn, Westchester, New Jersey, Connecticut and Eastern Long Island. CWB is experienced in historic renovation, residential architecture adapted for modern living and commercial spaces crafted for collaboration and productivity. CWB works with clients' needs, vision and budget to derive imaginative solutions. Part of that process includes identifying design challenges at the start of a project, so the finished work feels integrated and effortless. The firm's work in urban settings (where limited square footage presents unique demands) helped shape its core philosophy. The creation of spaces that are simultaneously simple and complex, elegant and modest, luxurious and understated is the basis of all CWB's work.

UXUS

Established in 2003, UXUS is a leading global strategic design consultancy delivering innovative consumer experiences for top multi-national brands. Regarded internationally as a thought leader, UXUS produces emotional and intelligent design exemplifying the principles of Brand Poetry: balancing creative excellence with commercial success. The award-winning team at UXUS consistently delivers the most unique and exciting solutions possible, creating noteworthy design and breaking through industry standards. UXUS has built a reputation for design excellence and innovation, combined with a comprehensive perspective into the world of consumer experiences and products. UXUS works with cutting-edge global brands in over 40 countries, including Qatar Luxury Group, Selfridges, Bloomingdale's, Chanel, InterContinental Hotels Group, Tate Modern, P&G, McDonald's and many other valued clients. In 2013, UXUS joined partnership with FutureBrand, a leading brand consultancy network with over 25 offices worldwide.

David Guerra Architecture and Interior Design

Founded in 2002 and located in Belo Horizonte, Minas Gerais, Brazil, the office David Guerra Architecture and Interior seek to combine creativity and functionality, always complying with the desire of the client in a singular and innovative way. The Office acts in the architecture and interior design fields, and its works range from the small scale of the object, to the big scale of public and institutional buildings. The team of architects and interns engage in extensive research to develop the projects, wherein the attention to the details and to the composition is always present. Numerous projects of David Guerra Architecture and Interior have been awarded several times and have been published extensively, both in national and international media.

Chango & Co.

Chango & Co. is a full-service New York interior design firm specializing in interior design, interior architecture, custom furniture design, audiovisual design and art curation & placement for high-end residences, restaurants & hotels. The firm is led by creative director Susana Simonpietri. Over the last decade, Susana has held the position of Senior Project Designer for some of the most prominent design firms in America, taking the lead designer and project manager roles for various international resorts, hotels, restaurants, and high-profile private residences. The "company" in Chango & Co. consists of the firm's creative & managing director, junior designers, Pratt Institute interns & a trusted team of contractors, custom furniture makers, upholsterers, a structural engineer & and on-staff art curator & assessor.

Nieto Design Group

Luxury Interior Design, Architecture & Space planning services that are individualized for each client to create unique environments that are designed for their lifestyle, tastes & preferences. Complete turn-key operations from conceptual to finished product including signed & sealed plans, permitting, 3-D renders, furniture & finish selection.

Leonidov and Partners

We tried to follow the idea of integration of the exterior into the interior. The House is situated in a nice village not far from Moscow. We tried to keep the feeling of wildness and natural beauty of the landscape, so we took a decision to reflect this beauty in the interior. The cozy environment has been created by strengthening of decorative effects and color through a smooth, harmonious transition of tones into a richer chord.

López Duplan Arquitectos

In architecture, spaces must communicate with the people who live them and transmit their lifestyle. Each project is a commitment to both interior and exterior that generates a language through all the elements that shape it. The result is an atmosphere that is present in an intimate communication between the architect and his client. Architect Claudia Lopez Duplan, has more than 20 years of experience in the development of residential, corporate and commercial projects. Her unique sensitivity and style have positioned her in the interior design scene; and she has specialized in residential renovation that gives the spaces a new identity. In her projects, lighting plays an important role and the use of indirect light, combined with variation of intensity, creates different atmospheres required for each space. Keeping the original architecture is one of the challenges faced in each project, besides the integration of all elements giving that result in a whole new image.